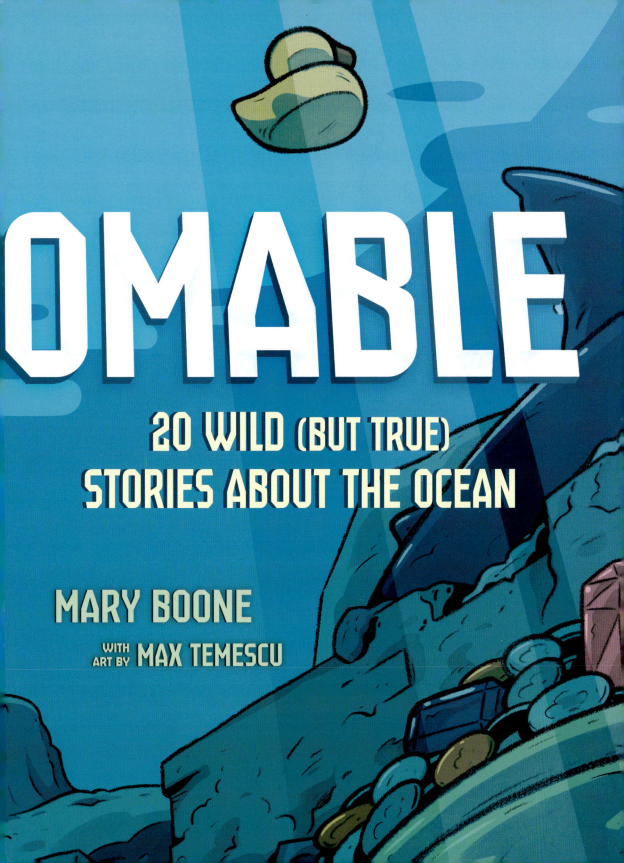

OMABLE

20 WILD (BUT TRUE) STORIES ABOUT THE OCEAN

MARY BOONE

WITH ART BY **MAX TEMESCU**

Bright Matter Books
An imprint of Random House Children's Books
A division of Penguin Random House LLC
1745 Broadway, New York, NY 10019
penguinrandomhouse.com
rhcbooks.com

Editor: Tom Russell
Designer: Jade Rector
Copy Editors: Maddy Stone and Alison Kolani
Managing Editor: Rebecca Vitkus
Production Manager: Patty Collins

Library of Congress Cataloging-in-Publication Data is available upon request.
ISBN 978-0-593-90475-6 (trade)—ISBN 978-0-593-90478-7 (lib. bdg.)—ISBN 978-0-593-90476-3 (ebook)

The text of this book is set in 12-point Narevik.

Manufactured in China
10 9 8 7 6 5 4 3 2 1

The authorized representative in the EU for product safety and compliance is Penguin Random House
Ireland, Morrison Chambers, 32 Nassau Street, Dublin D02 YH68, Ireland, https://eu-contact.penguin.ie.

FOR RON, WHO TAUGHT ME THAT
STORIES—LIKE OCEANS—HOLD HIDDEN TREASURES

CONTENTS

WELCOME TO
UNFATHOMABLE

MORE THAN 70 PERCENT OF EARTH'S SURFACE IS COVERED BY water. The oceans play an important role in bringing rain to farms and forests, as well as drinking water to our homes.

Even those who have never splashed in the waves or who live hundreds of miles from the nearest beach should know about our oceans. After all, they produce more than half the world's oxygen. We rely on them for transportation, jobs, recreation, and the production of food and medicine. Without the ocean, life on Earth would be impossible.

Our oceans are full of mysteries and surprises. In this book, you're going to discover some of the wildest, most wonderful things about the ocean. Some of the stories you're about to read are serious or sad, while others are downright funny.

It's true that some of these surprising tales may leave you questioning the things you thought you knew about science and nature. And that's all right, as long as they also get you wondering: *What else might be possible?*

CHAPTER 1

BLUBBER BLUNDER

ANYONE WHO HAS HAD A PET GOLDFISH DIE UNDERSTANDS THAT there are certain "acceptable" ways to bid it farewell. You might hold a family memorial service and bury the animal in a shoebox in your backyard. No yard and your family's in a hurry? Perhaps you could give it a funeral flush out to sea.

When a whale dies, flushing is not a possibility (even if you could find a giant toilet).

So what do you do when a dead forty-five-foot-long sperm whale washes up on the shore of your town? That was the question facing the residents of Florence, Oregon, back in 1970. It was early November when the whale washed ashore, and the eight-ton carcass, longer and heavier than a school bus, immediately began to attract attention. In the U.S., it is against the law to touch living or dead marine life on public beaches—plus it can carry all sorts of diseases. But that didn't stop people from driving hundreds of miles to this small West Coast town to see the massive mammal. A few visitors left bouquets of flowers to honor the deceased beast, while others posed for photos with it.

After a few days, though, the whale began to decay. Its skin began to dry and shrivel. Slimy whale juices pooled on the beach and gases began to accumulate in the animal's belly, filling the air with a putrid, stomach-turning odor. The aroma was sour, acrid, and unforgettable. It was as if you had dived headfirst into a pile of steamy garbage and the goo clung to the inside of your nostrils. Waiting for the whale to decompose on its own could take months, maybe even a year. Everyone agreed: The stink had to go, which meant the whale had to go. But no one knew exactly how to make that happen.

In Oregon, individual homeowners and businesses cannot own any part of the coastline. All the state's beaches are public. That was good news in the sense that no single family or business was responsible for whale disposal. But it also meant it was up to the state government to figure out what to do with the whale. In the end, the task of getting rid of the carcass fell to the Oregon Highway Division.

Whale removal is not an everyday task. Somewhere around 2,000 whales and dolphins wash up on beaches worldwide every year, a phenomenon called beaching or stranding. Most of those animals die, and many are left to rot in place along unpopulated stretches of coastline. That was not the case with the Florence whale. People lived nearby, businesses needed to operate, tourists wanted to visit, but none of that could happen if the stink stayed. Plans needed to be made—and quickly.

Initially, there was talk of hauling the carcass out to sea, but many worried that the tide would wash it back onto the beach. The whale could be moved to a more remote location, but it was already falling apart, and no one wanted to leave a trail of whale parts along the way. The whale *could* be buried, but what if waves washed away the sand and uncovered the corpse? Suggestions about cutting it up and burning it were immediately ruled out, over concerns both methods might upset spectators who had become emotionally attached to the dead whale. Plus, whales are considered sacred animals in many cultures—respectful disposal was paramount.

State officials were just about out of ideas when they got a phone call from the United States Navy. Their teams occasionally used

> SOMEWHERE AROUND 2,000 WHALES AND DOLPHINS WASH UP ON BEACHES WORLDWIDE EVERY YEAR, A PHENOMENON CALLED **BEACHING** OR **STRANDING**.

> WHALES ARE CONSIDERED SACRED ANIMALS IN MANY CULTURES— **RESPECTFUL DISPOSAL WAS PARAMOUNT**.

dynamite to break apart large rocks. It took some time, but eventually navy officials convinced the highway department that boulders and blubber were not all that different, and a quick resolution would hopefully lessen the pain for the public. So explosives it would be.

After three days of adding, subtracting, multiplying, and dividing, workers concluded they would need twenty cases—about a half ton—of dynamite to blow up the whale. They figured the explosion would be so powerful that only teeny, tiny whale morsels would be left. Those scraps could be eaten by seagulls, hawks, vultures, crabs, and other wildlife. The cycle of life would go on.

On the day of the big blast, the two-lane highway leading into Florence was busier than normal. News about the planned explosion had been broadcast on all the area's television and radio stations. Regardless of its macabre nature, this had become a must-see event. Spectators arrived hours ahead of time and brought along beach chairs, blankets, and cameras. They tossed footballs, flew kites, and built sandcastles while they waited. Many even packed fancy picnic lunches. Barriers kept folks a safe distance from the explosion, but crowds gathered atop nearby sand dunes in hopes of getting the best views. This was going to be a day to remember.

And it was. But not for the reasons you might imagine.

Reporters talked into microphones and TV cameras recorded the action as the countdown began. Oregon highway engineer

George Thornton had been put in charge of the operation. During one pre-blast interview he said: "Well, I'm confident that it'll work. The only thing is that we're just not sure exactly how much explosives it'll take to disintegrate this thing." Thornton's statement would come back to haunt him. For decades, everyone from explosives experts to comedians have criticized both the amount of dynamite used and its placement around the whale. But that's getting ahead of the story. When Thornton made his comments, no one yet knew how things would play out.

Minutes after his speech, though, the countdown began. Five, four, three, two, one. BOOM!

The explosion sent big chunks of gooey whale and sand flying more than 100 feet into the air. The blubber, which was supposed to soar toward the sea, instead started raining down on people who had gathered to watch. This was not the plan—not at all.

> "WELL, I'M CONFIDENT THAT IT'LL WORK. THE ONLY THING IS THAT WE'RE JUST **NOT SURE EXACTLY HOW MUCH EXPLOSIVES IT'LL TAKE** TO DISINTEGRATE THIS THING."
>
> —GEORGE THORNTON,
> OREGON HIGHWAY ENGINEER

Hoots and hollers had greeted the initial blast. But almost immediately, those cheers turned to screams of horror as a drizzle of chunky, smelly whale juice fell to the ground. People covered their noses and ran for shelter as larger hunks of whale began dropping from the sky. *Splat! Thunk! Thwat!*

Globs of whale, some as big as a refrigerator, thudded down. Some flew as far as 800 feet.

Paul Linnman was a TV reporter who covered the event. He remembers the chaos, with people running every which way to escape the flying pieces of whale meat.

"Explosions in the movies usually look like a blast of fire and smoke," Linnman later wrote in a book. "This one more resembled a mighty burst of tomato juice."

Rookie reporter Larry Bacon was covering the event for Springfield, Oregon's newspaper *The Register-Guard*. "All of a sudden there was this hundred-foot geyser of blood, blubber, and sand going up into the sky," Bacon recalled years after the blast. "It was like a blubber snowstorm with tiny particles of blubber floating down after the big chunks."

No one was hurt in the Florence whale explosion, but everyone on the beach was pelted with whale bits. One chunk of whale was so large it crushed the roof of a new car that had been parked a quarter mile from the blast site.

> "EXPLOSIONS IN THE MOVIES USUALLY LOOK LIKE A BLAST OF FIRE AND SMOKE. THIS ONE MORE RESEMBLED A MIGHTY **BURST OF TOMATO JUICE**."
>
> —PAUL LINNMAN,
> TV REPORTER

Sadly, the whale-covered crowd wasn't even the worst part. When the smoke, gunk, and guts settled, more than half of the enormous mammal remained on the beach. Plus, the scavengers that workers had counted on to gobble up the tiny bits were frightened off by the blast.

In the end, crews had to gather scattered chunks of whale and bury them along with the rest of the carcass. A few select fragments of the whale's skeleton were donated to the Siuslaw Pioneer Museum, where they are displayed along with other remembrances of local history.

The Oregon Department of Transportation considered the 1970 project a success and blamed the media for making it seem otherwise. Still, when forty-one sperm whales beached themselves on the same Florence beach a few years later, the idea of blowing them up was barely even discussed. Instead, the whales were buried right there, in the sand.

A FEW SELECT FRAGMENTS OF THE WHALE'S SKELETON WERE DONATED TO THE **SIUSLAW PIONEER MUSEUM**, WHERE THEY ARE DISPLAYED ALONG WITH OTHER REMEMBRANCES OF LOCAL HISTORY.

There are many in Florence who refuse to talk about the 1970 fiasco, bristling whenever a reporter revisits the event or videos of the blast gain traction on YouTube (twenty million views and counting). Still, others have chosen to embrace this unusual chapter in the area's history, selling commemorative T-shirts and posters, and dressing up in whale costumes for special events. Another sign of acceptance? When the city developed a small riverfront park in 2020—five decades after the blast—residents voted to name it Exploding Whale Memorial Park.

The whale is gone. The stench is gone. But this coastal town's connection to one whale of a tale will live on forever.

CHAPTER 2
THE *JUNK*

"HONEY, WE'RE SINKING."

Those were the last words Anna Cummins wanted to hear from her then fiancé, Marcus Eriksen, when he called her via satellite phone on June 7, 2008.

Eriksen and fellow environmental activist Joel Paschal were aboard a raft called *Junk*. The boat was constructed of make-shift pontoons fashioned from old fishing nets filled with 15,000 two-liter plastic bottles. Its cabin was a discarded Cessna aircraft fuselage, and twenty-five broken sailboat masts formed its deck. It was all, quite literally, *junk*.

The duo was attempting to sail 2,600 miles, from Southern

JUNK WAS CONSTRUCTED OF MAKESHIFT PONTOONS FASHIONED FROM OLD FISHING NETS FILLED WITH **15,000** TWO-LITER **PLASTIC BOTTLES**.

California to Honolulu, Hawaii, with the hope of building awareness about marine plastic pollution. Cummins, who'd been heavily involved in planning and building the raft, was in California, serving as land-based mission control. She kept an eye on the weather, headed up fundraising, answered calls from reporters, updated the group's blog, and handled any problems as they arose.

Sinking was *definitely* a problem—a very big problem—and one that probably could have been avoided.

When preparing for their trip, Cummins, Eriksen, and Paschal had asked schoolchildren from all over Southern California to help collect and clean bottles for their mission. At one point a young girl asked the team: "Aren't you going to glue the tops on those bottles?"

"Nah, hand-tight is fine," Eriksen told her. "I've done it before."

Flash forward to day three of the raft's voyage: Ocean waves were loosening the bottles' grooved caps. Bottles were filling with water and the boat was sinking. Standing on the deck, Eriksen and Paschal both found themselves ankle-deep in water.

The eco-adventurers imagined how their story might end. Perhaps they'd have to be towed back to shore and donors would ask for refunds. Maybe they would have to spend the rest of their lives explaining that, yes, they were the guys who foolishly thought they could sail on a boat made of trash. Or—even worse—what if they drowned and the raft sank, further polluting the very ocean they were trying to save?

Fortunately, the pity party didn't last long. Back on land, Cummins jotted down a list of materials needed and quickly recruited a crew to help repair the raft. Within twelve hours, she and six volunteers were aboard a boat, with fifty tubes of glue, speeding toward *Junk*. They spent the next fourteen hours bailing water, draining bottles, and gluing caps on—all while still afloat.

Luckily, those repairs held, but it was hardly the only challenge the team faced during its eighty-eight-day journey. The boat was constantly falling apart. Storms sent waves surging onto the deck. Clothes and sleeping bags were soaked. There were near misses with other ships and ragged, rocky coasts. Weeks without much wind meant travel for the motorless raft was slow. Living for three months in such a small space left both men restless and irritable.

Because they'd originally thought the trip would take only half the time it did, food supplies had to be rationed. Eriksen and Paschal fished but quickly realized ocean life was not as plentiful as it had once been. When fish were caught, they were savored. The meat from a mahi-mahi, for example, was divided into several meals. Parts of it were sautéed with garlic, while other bits were dried and eaten as jerky. Its skeleton was hung to dry, for nibbling on later.

Another catch, later in their journey, was a real eye-opener. The two famished men were initially elated when Paschal caught a good-sized rainbow runner. But, as Eriksen cleaned the fish and prepared to cook it, he discovered its stomach was unusually hard. The men's appetites disappeared when they discovered the fish's stomach was filled with plastic.

"It shouldn't have been a surprise," says Cummins. "I mean, the whole purpose of the voyage was to build awareness and help create a movement to save our oceans from plastic pollution. But still, this fish wasn't caught along the coast, where you might expect waters to be polluted. It was caught in the middle of nowhere. It was startling—even to us—to see how bad and how prevalent the pollution was."

On August 28, 2008, three months after launching, Eriksen and Paschal landed safely in Honolulu, Hawaii. They were dirty and hungry, but excited to meet with locals and journalists who greeted them onshore. The adventurers' next challenge was to educate people about the pollution they saw in the middle of the Pacific Ocean.

Junk wasn't Eriksen's first vessel made of trash. Previous adventures had taken him down

"THIS FISH WASN'T CAUGHT ALONG THE COAST, WHERE YOU MIGHT EXPECT WATERS TO BE POLLUTED. IT WAS CAUGHT IN THE MIDDLE OF NOWHERE."

—ANNA CUMMINS, ENVIRONMENTAL ACTIVIST

the Chattahoochee, Los Angeles, Mississippi, and Potomac Rivers, and along the California and Alaska coasts aboard crafts named *Spastic Plastic*, *Cola Kayak*, *Bottle Rocket*, *Potomac Attack*, *Fluke*, and *Plastic Poison*. He and Cummins have long conducted research related to marine conservation, sustainability, and plastics pollution.

In the years since *Junk*'s Pacific Ocean expedition, Eriksen and Cummins got married and founded the 5 Gyres Institute. (Gyres are large systems of rotating ocean currents—or spots in the ocean where the waves spiral, causing debris to collect in one location. Over the years, garbage has accumulated in the world's five subtropical gyres: the North and South Pacific Gyres, the North and South Atlantic Gyres, and the Indian Ocean Gyre.)

The 5 Gyres Institute focuses on leading expeditions, raising awareness, and working with

PREVIOUS TRASH VESSELS

SPASTIC PLASTIC

COLA KAYAK

BOTTLE ROCKET

POTOMAC ATTACK

FLUKE

PLASTIC POISON

JUNK

GYRES ARE LARGE SYSTEMS OF ROTATING OCEAN CURRENTS— OR SPOTS IN THE OCEAN WHERE THE WAVES SPIRAL, CAUSING DEBRIS TO COLLECT.

communities to enact changes that benefit the environment. During a 5 Gyres research trip in 2015, for example, they found plastic microbeads—tiny round pieces of plastic used in toothpaste and face scrubs, among other products—in the Great Lakes. That research started a movement. There was testimony in front of the United States Congress. Eventually, US President Barack Obama signed the Microbead-Free Waters Act. The law went into effect in 2018. It prohibits the manufacturing, packaging, and distribution of personal care products that contain plastic microbeads. Eliminating single-use plastic bags, reducing packaging, and improving recycling programs are among other 5 Gyres priorities.

Will the team take on another *Junk*-sized environmental challenge?

"It's hard to say," says Cummins. "Sometimes you have to do something risky and out of the box to get eyes and ears on a problem. That's a big message we share with kids when we visit schools. Sometimes you have to jump up and down and wave your arms to get people to pay attention to what you're saying. For us, *Junk* was a good way to get people to tune in."

Building a raft out of trash and sailing on the open ocean is obviously not something everyone

THE **MICROBEAD-FREE WATERS ACT** PROHIBITS THE MANUFACTURING, PACKAGING, AND DISTRIBUTION OF PERSONAL CARE PRODUCTS THAT CONTAIN PLASTIC MICROBEADS.

should do. But Cummins notes: "Every single person has a voice and the power to make little changes, in our families or households. We've met kids who've fought for changes in their schools, who've gotten petitions signed, and attended city council meetings. In their own way, they're spreading the word: This isn't an ocean issue, it's a public health issue."

CHAPTER 3
MY WRESTLING OPPONENT HAS HOW MANY ARMS?

GARY KEFFLER'S WRESTLING OPPONENT WRAPPED ONE ARM around his belly and another around his neck. A third arm grabbed for his legs.

Um. Wait. A third arm?

Yes. And a fourth, fifth, and sixth—and two legs, too.

This was no ordinary wrestling match, and Keffler was facing no ordinary opponent. This was human versus cephalopod, man versus mollusk. The year was 1963, and a then sixteen-year-old Keffler was part of a local diving club called the Puget Sound Mudsharks. To attract members, the club held group dives and spearfishing competitions. They also organized and competed in the World Octopus Wrestling Championships, an event held in Tacoma, Washington, in the late 1950s and 1960s.

Tacoma is located along the Puget Sound, an inlet of the Pacific Ocean and part of the Salish Sea. In July 1940, a poorly engineered suspension bridge was opened across the Tacoma Narrows strait of the Puget Sound. The bridge was short-lived, dramatically collapsing four months later. The remains of the bridge are still on the bottom of Puget Sound. But that's not the end of the bridge's story. Marine life grew around the steel frame and concrete slabs, forming an artificial reef that became home to a large population of giant Pacific octopuses. The octopuses were there. The divers were there. And so the competition began.

Teams of three divers—both men and women—entered the chilly waters and had two hours to locate octopuses and try to bring them to the surface. Once the sea creatures were out of the water, they were weighed. There were two categories of competition: one if you used scuba gear and air tanks and one if you free dived—or swam without using oxygen. As the bounty was weighed, teams were awarded two points per pound of octopus they collected free diving and one point per pound if the divers wore gear. The team with the most points won the trophy. Awards also went to the best rookie team and the team that brought in the smallest octopus.

The contest had two primary rules: The octopuses had to be taken up by hand—"no spears, lassos, or handcuffs." And each octopus had to arrive at the surface with its skin unscathed.

Finding the octopuses was often the most difficult part of the challenge. They dwell in dark dens, and because their bodies are entirely compressible, they can fit into small crevices and through the tiniest of holes. They can also change color and pattern to blend in with even the most intricate plants, rocks, or coral, making them pros at hide-and-seek.

Once an octopus was located by a keen-eyed contestant, a diver would grab the center of its body, where the arms attach to the head. The "wrestling" aspect of the competition came not because the octopuses were especially strong, but because they were slippery, wily, and difficult to wrangle. While one team member engaged the octopus, the others worked to keep the first diver from being wrapped up in wandering appendages. Octopuses have six front arms and two back legs, after all.

> THE OCTOPUSES HAD TO BE TAKEN UP BY HAND. **AND EACH OCTOPUS HAD TO ARRIVE** AT THE SURFACE WITH ITS SKIN **UNSCATHED**.

"It wasn't like they were going to squeeze you to death—they don't have that kind of strength," said Keffler. "But you had to be careful about them grabbing your face mask or pulling at your air hose."

In one of the last contests, a crowd of 5,000 people gathered along the shores of Titlow Beach one cool April morning to watch more than 100 divers take the plunge. The unusual event was covered by newspapers, magazines, and national television broadcasts, including ABC's *Wide World of Sports* and NBC's

Sports in Action. Despite the huge audience, octopus wrestling was not much of a spectator sport. The real action took place thirty to sixty feet underwater.

"When you grabbed one out of its den, that's when the real rassling occurred," said Keffler. "They've got pretty good suction. But once you give 'em a pull, the suction cups *pop, pop, pop*, one right after another."

In truth, that popping could go on for a while. Depending on their size, octopuses can have 100 to 280 suckers on each arm. They use those suckers to grab their dinner—often shrimp, fish, lobsters, and clams—and rely on their sharp beak-like mouths to break open shells and fight off predators. While it's not common, octopuses have been known to overpower and kill larger adversaries—including sharks. All octopuses are venomous; the giant Pacific octopus's venom isn't dangerous to humans, but it can be used to stun or paralyze prey.

These creatures, with their bulbous heads, grow bigger and live longer than any other species of octopus. The largest ever found measured 30 feet across and weighed more than 600 pounds; the majority are significantly smaller, weighing between 20 and 110 pounds. Most years the biggest octopus hauled in during the

> THE LARGEST GIANT PACIFIC OCTOPUS EVER FOUND MEASURED **30 FEET ACROSS** AND WEIGHED MORE THAN **600 POUNDS**.

Tacoma contest weighed around 60 pounds. In 1965, the Seattle Public Aquarium offered a $1,000 prize to any diver who brought up an octopus exceeding what was then the contest record of 89 pounds.

The aquarium provided holding tanks so those attending the competition could view the captured octopuses. Some of the mollusks were later put on display at the aquarium. Others were cooked and eaten. The rest were fitted with United States Fish and Wildlife Service tags and returned to the ocean, allowing their habits and migration patterns to be studied.

Keffler, who ran a chain of dive shops for many years, said he never feared his oceanic wrestling opponents. He was, after all, an experienced diver. He took second place in the World Spearfishing Championships in Brazil in 1963 and was a stunt double for actor Lloyd Bridges, who played scuba diver Mike Nelson from 1958 to 1961 in the action-adventure television series *Sea Hunt*.

The allure of octopus wrestling faded by the mid-1960s. Fewer divers showed up to compete, and crowds grew smaller. People were becoming more aware of animal cruelty and the need to care for nature. The final blow was dealt in 1976, when the state of Washington passed a law making it illegal to capture or harass an octopus; although, with the proper permits, they could still be fished

IN 1976, THE STATE OF WASHINGTON PASSED A LAW MAKING IT **ILLEGAL** TO CAPTURE OR HARASS AN OCTOPUS.

for food purposes. Over the years, restrictions have tightened even more. By 2010, octopuses in the waters around Washington could *only* be harvested if they were brought in without the use of knives or spears. And giant Pacific octopus hunting is now completely banned across seven different Puget Sound sites.

Despite taking home his share of trophies, Keffler admits the battle between humans and octopuses was never fair. In 1971, Keffler and his wife, Joanne Duffy, helped marine explorer Jacques Cousteau with the filming of "Octopus, Octopus," an episode of his documentary show *The Undersea World of Jacques Cousteau*. The feature dispelled myths portraying octopuses as vicious monsters and played a role in changing attitudes—Keffler's included—about the animals.

"Back then, we rassled octopus because they were there. . . . It was something to do," Keffler said. "We didn't really know how smart they are," he continued. "They're really beautiful, graceful, interesting creatures."

Octopuses have been the subject of a good deal of research over the past few decades. They are considered by many to be the most intelligent and evolved of all invertebrates. They can solve puzzles, differentiate between shapes, and recognize faces, and they possess both long-term and short-term memory. Divers still seek them out, not to wrestle but rather to study, admire, photograph, and revere.

CHAPTER 4

DOLPHINS TO THE RESCUE

THINGS DID NOT LOOK GOOD FOR TODD ENDRIS. HE AND SOME friends had been surfing off the coast of California on a chilly August morning in 2008, when—out of nowhere—it happened. A 3,000-pound great white shark lunged at him and tossed him into the air. Rattled but unhurt, Endris swam back to his surfboard and paddled as fast as he could toward shore.

He wasn't fast enough.

The shark returned, this time biting down hard, pulling Endris under the water and shaking him. The animal's razor-sharp teeth shredded the skin on Endris's back. Shocked and bleeding, Endris clung to his surfboard. As onlookers watched helplessly from the shore, a group of unlikely rescuers swam to Endris: a pod of bottle-nose dolphins.

The dolphins had been playing in the same waves that attracted Endris and his fellow surfers. Following the initial attack, they formed a protective ring around the injured surfer. They slapped their tails against the water in what eyewitnesses called an obvious attempt to scare off the shark. It worked, for a while.

The shark eventually broke through the dolphins' defensive barrier and attacked Endris again. Its powerful jaws clamped down on his leg, slicing through the muscle on his right thigh. And, as if all this wasn't enough, the shark wouldn't let go. Endris feared he might die, but the hero dolphins continue to wage war against the shark.

"While all of this is happening, there are dolphins jumping over my head, literally jumping over me, doing flips over me, tail slapping the water inches away from my head," Endris recalled in a National Geographic video segment about the event. Meanwhile, under the water, dolphins continued to call—whether to one another or to the injured surfer or both, no one is sure. But eventually the shark retreated and, with the help of a surfer pal, Endris made it safely back to shore.

"If one of the dolphins had been just an inch or two off, it would have knocked me unconscious easily and I would have died right there," said Endris. "They knew what they were doing. And they were literally working to save me. They were keeping the shark from coming back."

Even with the dolphins' aid, Endris's injuries were significant. The average adult human body contains around 10.5 pints of blood; Endris lost six pints. His puncture wounds and cuts required 500 stitches and 200 staples. He underwent months of physical therapy to regain strength and repair the muscle damage suffered during the attack.

"Dolphins are just the most amazing things," said Endris. "They're extremely intelligent and they have emotion. That day I was truly lucky. Those dolphins definitely saved my life."

Endris is not the only person dolphins have rescued. The Whale and Dolphin Conservation Society says accounts of the marine mammals' heroics go back to ancient times. There's even a popular myth that tells of Taras, the son of the Greek god Poseidon, who controlled the seas. According to the story, Taras was saved from a shipwreck by a dolphin that carried him back to dry land.

TARAS, THE SON OF THE GREEK GOD POSEIDON, WAS SAVED FROM A SHIPWRECK BY A DOLPHIN WHO CARRIED HIM BACK TO DRY LAND.

For many years, people assumed Taras's story was just another fictional tale. But the more people have learned about dolphins, the more likely it appears that this myth may be based on a real event. Over the years, there have been many accounts of dolphins saving or protecting humans.

In 2004, four lifeguards were swimming off the northern coast of New Zealand when seven bottlenose dolphins swam quickly toward them and herded the swimmers together.

"They were behaving really weird, turning tight circles on us, and slapping the water with their tails," Rob Howe told *The New Zealand Herald* after the event.

Howe had swum with dolphins before. At first, he wasn't sure why these animals were acting so strangely. Then he saw it: a ten-foot-long great white shark. The dolphins, he realized, were creating a screen to protect the swimmers. The diving, flipping, and water slapping continued for forty minutes, until the shark left the area.

In 2014, Adam Walker was crossing the Cook Strait, which separates the two islands of New Zealand, in a long-distance charity swim. He was raising money for dolphin conservation. The sight of a great white shark lurking in the water beneath him got his heart racing. But then the very animals he was working to save gathered around him. The circle of ten dolphins swam with him for more than an hour, until the shark lost interest. "I'd like to think they were protecting me and guiding me home," Walker wrote on social media. "This swim will stay with me forever."

THE CIRCLE OF **TEN DOLPHINS SWAM WITH WALKER** FOR MORE THAN AN HOUR, UNTIL THE SHARK LOST INTEREST.

A lost swimmer was found off the coast of Ireland in August 2021, thanks to some helpful dolphins. The swimmer had been missing for nearly twelve hours, despite rescuers actively searching for him. Finbarr O'Connell, the rescuer credited with finding the swimmer, told *The Irish Independent* that there were a lot of dolphins around the man when he was located. "Maybe they helped him in some way or another. Who knows?" he said.

Researchers are quick to agree that dolphins are very intelligent—much like octopuses. These animals are known for their problem-solving, innovation, and ability to learn quickly. However, *why* dolphins rescue people is something experts do not see eye to eye on.

DOLPHINS AND OCTOPUSES ARE VERY INTELLIGENT. THEY ARE KNOWN FOR THEIR PROBLEM-SOLVING, INNOVATION, AND ABILITY TO LEARN QUICKLY.

There are researchers who believe it's a basic instinct for dolphins to help other dolphins. Helping a human is just a small step from that impulse. Some believe the dolphins are trying to protect themselves or their young from danger. Occasionally, a human may be nearby, so saving them is a bonus, but it's not their main mission.

Still others think dolphins help humans simply because they're curious. They circle around surfers and swimmers because they want to learn more about them. A shark is also nearby? Maybe that's just a coincidence.

Richard C. Connor is a biologist who has been studying dolphins for more than forty years. He says the stories about dolphins rescuing humans are fascinating, but until he sees it firsthand, he's skeptical. He asserts that dolphins are socially complex and their alliances shift. "That includes doing things that, from our point of view, are nice and not so nice," he said.

Marine biologist and shark conservationist Tom Hird's stance on the matter is similar to Connor's. "Being saved by another animal is certainly a very romantic notion," he told National Geographic. But, he adds, that doesn't mean it really happens.

Lori Marino, an expert in animal behavior and intelligence, likes to give dolphins a little more credit. "Dolphins possess large, complex brains," she says. "Scientists have concluded that they're capable of sensing other creatures' emotions. I don't think it's a stretch at all to believe that dolphins can sense when a human is in trouble and respond accordingly.

"If I was in the ocean and I was in trouble," she continued, "I'd hope there were dolphins nearby to provide aid. In my mind, I'd rather take my chances that they'd help me than to be out there all by myself."

CHAPTER 5

BATTALION OF BATH TOYS

JANUARY 10, 1992: THE SHIP *EVER LAUREL* WAS HALFWAY through its journey from Hong Kong to Tacoma, Washington, when a storm hit.

The waters south of Alaska's Aleutian Islands are known for being rough and wild. This day was no exception. Hurricane-force winds roared, tossing the ship to and fro, while thirty-six-foot waves crashed against the deck. The vessel jerked so violently that twelve large shipping containers broke loose. They slid and tumbled from the deck. One of the containers broke open as it fell into the ocean. No one could have imagined the impact its contents would have on science for years to come.

That metal box contained 28,800 plastic bath toys: yellow ducks, red beavers, blue turtles, and green frogs. The toys, named Friendly Floatees, were packed into 7,200 cartons. And when salt water dissolved the packaging, a battalion of bath toys was set free. As adorable as these little bath toys were, they were still flotsam. That's the fancy word for stuff that's unintentionally thrown overboard, often as the result of a shipwreck or a collision. What might initially be seen as a tragic dumping of more plastic into the ocean instead became a powerful way for researchers to study currents.

You see, ten months later, those toys started to wash up along the Alaskan coast nearly 2,000 miles from where they first plummeted into the sea. At first, it was only ten toys, but over the next year, another 400 duckies, beavers, turtles, and frogs were found along the eastern coast of the Gulf of Alaska.

In 1995, a Friendly Floatee washed ashore in Washington State. In 1996, another was found 4,000 miles away on a Hawaiian beach. The toys were venturing far and wide.

Along the way, the duckies and friends captured the attention of oceanographer Curtis Ebbesmeyer, who had spent his career studying ocean currents.

From 1965 to 1974, Ebbesmeyer worked for Mobil Oil Corp, where he tracked icebergs off the coast of Newfoundland, Canada. If an iceberg got too close to one of the company's oil-drilling rigs, Ebbesmeyer helped figure out how to tow it away. He later worked for a consulting firm, where he monitored ocean turbulence deep below the surface and became an expert on eddies. Those are small, spinning currents. He helped track the 1989 Exxon *Valdez* oil spill in Prince William Sound, Alaska, and he has studied the ways in which water movement affects everything from sewage spills to migrating salmon.

In the 1990s, most ocean scientists used drift bottles to learn what they could about currents. Scientists would seal messages into glass containers and throw them from ships or planes in specific locations. They were allowed to release a maximum of 1,000 bottles per investigation. As some of those bottles washed ashore, their paths could be mapped. But our oceans are vast and unpredictable. Only about 2 percent of drift items are ever recovered. So, out of 1,000 released bottles, scientists were only able to gather data from about twenty of them.

Drift bottles were eventually replaced by drift cards. Drift cards are small, eco-friendly wooden cards that float along the water's surface. They are moved by currents, wind, and tides. Each card is coded so researchers can identify where it was deployed. Drift cards are still used by some groups today. Currents are also monitored through the use of aerial photography, satellite networks, GPS-enabled sensors, and powerful computers.

DRIFT CARDS ARE TOOLS SCIENTISTS USE TO STUDY CURRENTS, WIND, AND TIDES.

These days, no one would intentionally toss tens of thousands of pieces of plastic into the ocean—even in the name of science.

But thanks to the *Ever Laurel*, the bath toys were already in the water, and researchers knew exactly where they fell in. Plus, there were so many of them that at least 600 of them should eventually wash ashore. The potential for learning more about ocean currents was huge.

Friendly Floatees executives felt bad about the spill, and they wanted to help science. To do their part, the company began offering beachcombers small rewards if they reported the location where a bath toy had been found. That information was entered into a computer model that factored in wind data and other variables to map the toys' suspected routes. This data helped researchers better understand the five major ocean gyres. (Yep, the same gyres Marcus Eriksen and his team researched in chapter two.) Scientists knew the gyres existed, but until they began following the bath toys' movements, no one knew it takes about three years to complete a single circuit of the North Pacific Gyre. Experts tracking the toys estimate as many as 2,000 of them are still stuck in the currents, going round and round and round. A smaller group of the bath toys split off into the South Pacific Gyre—a gyre that touches the equator.

In 2003, after a long while with no Friendly Floatee sightings reported, the toys began showing up again, this time in the Atlantic Ocean. It's believed that the toys had been pushed northward through the Bering Strait, which acts as a gateway between the

Pacific and the Arctic Oceans. The toys likely became frozen in Arctic pack ice, a large mass of ice pieces that have been driven together by wind or currents. The ice moved them eastward before melting as it reached the northern Atlantic Ocean. A duck was found in Maine. A frog was discovered on the shores of Scotland. Other toys were found along the coasts of England and Ireland.

ARCTIC PACK ICE IS A LARGE MASS OF ICE PIECES THAT HAVE BEEN DRIVEN TOGETHER BY WIND OR CURRENTS.

While all this ducky data has helped scientists better understand ocean movement, it's hardly the only spillage being studied.

In May 1990, a cargo ship called the *Hansa Carrier* was headed from South Korea to the United States when a storm hit. It lost twenty-one shipping containers; five of them were filled with nearly 79,000 Nike shoes. Identifying numbers stamped onto the shoes later revealed that four of the Nike containers had broken open, releasing 61,280 shoes. The pairs were not laced together, so each shoe became a separate piece of flotsam.

The shoes drifted in the Pacific for months. But here's where it got weird—the shape of the shoes determined which way they drifted. For the most part, the right shoes headed northeast into the Alaska current. The left shoes joined the California current. The spill was a public relations nightmare for Nike, but for beachcombers it created the opportunity to form a network to match pairs, clean them up, and sell them. The sneakers also provided new insight for those studying ocean currents.

Both the Friendly Floatees and the Nikes provided something that's rare with flotsam. "[We had] both point A, when and where an object starts to drift, and point B, when and where it washes up," Ebbesmeyer wrote in *Flotsametrics and the Floating World*, which he coauthored with Eric Scigliano. That data was fed into computer models that were able to forecast current movements.

These notable bits of ocean debris also caught the attention of media. The role they played in tracking currents has been featured in books, and many schoolteachers still lead students through lessons on the famous flotsam.

"It's deeply gratifying to think that some kids might dedicate their lives to studying and protecting the oceans after getting an inspirational nudge from the saga of the telltale sneakers," Ebbesmeyer wrote.

What else has accidentally fallen into the ocean? Everything from containers of flat-screen TVs to hockey gloves, and motorbikes to printer cartridges. According to the World Shipping Council, as many as 1,382 shipping containers are lost at sea each year. Those steel containers vary in size; the largest are typically 40 feet long and can fit the contents of a five-to-seven-bedroom house. Unfortunately, most of the items from these containers don't help scientists learn about currents. Shipping companies are not required to report lost cargo, so the public often only learns about it when the items start washing ashore and attract the attention of beachcombers, local authorities, and the media.

A load of lost cargo that caught the spotlight came in the shape of Garfield the cat telephones. Dozens of news articles were written about the phones for two obvious reasons: First, the phones were funny. They were bright orange and the cartoon cat's eyes would open wide when the receiver was

ACCORDING TO THE WORLD SHIPPING COUNCIL, AS MANY AS **1,382 SHIPPING CONTAINERS ARE LOST AT SEA** EACH YEAR.

picked up. They were very trendy in the 1980s and the sight of one wrapped in seaweed and washed up on a beach was oddly amusing. Secondly, the devices had been washing onto the beaches along the northwestern coast of France for thirty years. And, even after all that time, no one could figure out why the phones, receivers, and their coiled orange cords were appearing year after year.

In 2019, a local farmer solved the mystery. He said that he and his brother had made a discovery in the mid-1980s. They'd found a metal shipping container filled with the phones deep inside a sea cave. High tides make it impossible to enter the cave most of the year. Environmentalists were hoping that when it was finally safe to enter, they'd find the container still brimming with phones—debris they could prevent from entering the ocean. But that wasn't the case. The container was empty, and bits of Garfield phones were perched on nearby rocks. There was nothing left that might help researchers deepen their under- standing of the ocean—just another lost container, its contents turned to litter.

CHAPTER 6
STINGRAYS

ON A STEAMY DAY IN AUGUST 2023, KRISTIE CATAFFO-O'BRIEN and her husband drove to Bahia Beach, along Florida's Gulf Coast. Looking to cool off, the thirty-eight-year-old nurse leaned back to wet her hair in some shallow water. That's when she felt a sharp stabbing pain in her upper back. She had been stung by a stingray.

The ray's razor-sharp barb sliced through Cataffo-O'Brien's skin and wedged itself three inches deep in muscle, barely missing her lungs. Even worse? When she sat up, the live ray was dangling from her back, still flopping.

"It was quite terrifying," she told reporters in the days following the incident. She later wrote on social media: "I have never been so scared and actually thought I was going to die."

Paramedics rushed to the scene and used scissors to cut the stingray's barb. A surgeon removed the imbedded portion of the barb, and medications were prescribed to counter the stingray's venom. Cataffo-O'Brien spent a week in the hospital, and it took months for her to fully recover.

Cataffo-O'Brien's experience was not typical. Stingrays are not usually aggressive. They strike out with their barbs only when threatened. The barbs, also known as spinal blades or stingers, are made of cartilage, the same stuff human ears and noses are made of. The barb is covered by a thin layer of skin and mucus.

If you know anything about stingrays, you likely know they spend a lot of time foraging or resting on the seafloor. They hide from predators by burying themselves in the sand. It stands to reason then that they may feel threatened when someone steps on or—as was the case for Cataffo-O'Brien—sits on them.

STINGRAY BARBS, ALSO KNOWN AS SPINAL BLADES OR STINGERS, ARE MADE OF CARTILAGE.

For this reason, most stingray injuries reported each year are to the feet or legs; those are the body parts most likely to come in contact with a hidden ray. Stings to the lower body probably won't kill you, but they still hurt quite badly. The pain, which may be accompanied by sweating or muscle cramps, usually goes away within six to forty-eight hours.

Dr. Christopher G. Lowe is a marine biologist and director of the Shark Lab at California State University, Long Beach. Researchers at the Shark Lab also study rays because the two types of fish are closely related. One hazard of his job? Lowe has been stung a couple dozen times. It happens. In fact, Shark Lab estimates that 10,000 beachgoers are injured by stingrays each year in Southern California alone.

"It definitely hurts," says Lowe. "But size matters. The amount of pain you feel is often related to the size of the ray you encounter."

Round rays (*Urobatis halleri*) are common along Southern California beaches. Their maximum body length is 22 inches (56 cm), and their barbs are no more than 1.5 inches (3.8 cm) long.

> SHARK LAB ESTIMATES THAT **10,000 BEACHGOERS** ARE INJURED BY STINGRAYS EACH YEAR IN SOUTHERN CALIFORNIA ALONE.

"If you step on a round ray, they'll shoot their barb straight up like a nail or they'll swing it back and forth," says Lowe. "Those little barbs can still cut through skin and venom can get into your system, but they're nothing compared to some of the bigger stingrays."

How big are those "bigger rays"? The smalleye stingray (*Megatrygon microps*) is among the largest. These rare beasts are found in the Indian and west Pacific Oceans. They can grow up to 10 feet (3 m) long with a spine as big around as an adult's forearm.

Experts stress that stings from even the biggest rays are rarely deadly. No one seems to know for certain how many people have died from stingray attacks; estimates range from seventeen to thirty worldwide. That's altogether—throughout history, or for at least as long as people have been keeping track of these things. By comparison, an average of seventy-two people die each year because of hornet, wasp, or bee stings.

Which is why fatal—and nearly fatal—stingray attacks often make the news. It's also why Cataffo-O'Brien, still wincing in pain, immediately thought back to the most famous stingray victim of modern times: zookeeper, conservationist, and television host Steve Irwin.

Irwin was an Australian-born environmentalist whose high energy and boundless enthusiasm for nature propelled him to global stardom. He hosted several television series, most notably *The Crocodile Hunter*. Audiences were spellbound by Irwin's dangerous on-screen encounters with animals. He thought nothing of poking at venomous snakes, lassoing a tiger shark, or scurrying up a tree to escape an angry Komodo dragon—all while sharing knowledge about the wild creatures. Dressed in his trademark khaki shirt and shorts, Irwin would respond to a quickly advancing snapping crocodile with a spirited "Crikey!"

In September 2006, Irwin had a day off from filming a new TV series called *Ocean's Deadliest* in the waters near the coast of Queensland, Australia. Not one to sit still, he convinced his diving partner and camera operator Justin Lyons to join him in the search for interesting sea life they could film for another project. The two jumped in a boat and had only been motoring for a few minutes when they spied a bull ray (*Aetomylaeus bovinus*). This large species of stingray is found in warm water, often along the coasts of Europe and Africa.

"We'd swum with stingrays many times before," Lyons recalled in a 2014 television interview. "This one was extraordinarily large, he was eight feet wide, a massive bull ray. It was very impressive."

THE BULL RAY (*AETOMYLAEUS BOVINUS*) IS A LARGE SPECIES OF STINGRAY FOUND IN WARM WATER, OFTEN ALONG THE COASTS OF EUROPE AND AFRICA.

Irwin and Lyons climbed into the chest-deep water. They'd been shooting for a few minutes, then stopped to plan out what was to be their last bit of video for the day: With the ray in the foreground, Irwin would swim up from behind the animal.

"I thought, 'This is going to be a great shot,'" Lyons recalled. Then, suddenly, the stingray reared up. "It started stabbing wildly with its tail, hundreds of strikes within a few seconds."

The cameraman was focused on the ray as it swam away. He didn't realize the severity of Irwin's injuries until he saw the water darkening with blood.

Lyons, fearful the blood would attract sharks, quickly pulled Irwin from the water. "We hoped for a miracle," he said. "I literally did CPR on him for over an hour." The barb pierced Irwin's chest. Jabs to the arms or legs probably could have been survived, but Irwin went into cardiac arrest and was pronounced dead following the attack. Lyons later speculated that the ray may have thought Irwin's shadow was a predatory shark.

Irwin's death shocked the world. Following his passing, the Australia Zoo broadcast a public memorial service, and an estimated 300 million people tuned in. Irwin's death also begged the question, if someone with as much experience as he had could experience grave danger, should everyone fear wildlife?

The truth is accidents like the one that killed Irwin are unusual. Even more uncommon? Stingray-related injuries and deaths that occur *out* of the water.

Just six weeks after Irwin's death, James Bertakis was on his boat, fishing near his home in Lighthouse Point, Florida. Out of nowhere, a stingray sprang from the water and landed in his lap. Hold everything! Stingrays can fly? Not exactly. But it may be surprising to know they *can* jump out of the water to escape predators, attract mates, or remove parasites. The stress of these out-of-water experiences have even resulted in stingrays giving birth in midair.

Bertakis was definitely surprised. He wasn't immediately sure what had happened. He grabbed the ray and pulled. The ray's barb broke off in his chest.

"I didn't know what a barb was," he told a reporter in the days following the incident. "I felt a little pain. I didn't know what it was. I just turned around and threw the stingray back in the boat."

STINGRAYS CAN JUMP OUT OF THE WATER TO ESCAPE PREDATORS, ATTRACT MATES, OR REMOVE PARASITES.

Bleeding and lightheaded, Bertakis steered the boat to shore, climbed onto his dock, and asked friends to call 911. During a six-hour surgery, doctors removed the two-and-a-half-inch barb and repaired his damaged heart. A day later, his bruised spleen was removed.

Bertakis spent two months in the hospital, slowly rebuilding his strength. Shortly after he was discharged, he returned to his boat, figuring the odds of having another ray land in his lap were extremely low.

Michigan resident Judy Kay Zagorski's stingray encounter was both unpredictable and deadly. She was vacationing with family in the Florida Keys in spring 2008. Zagorski was sitting in the front seat of a boat going 25 miles (40 km) per hour, when a 75-pound

(34-kg) spotted eagle ray (*Aetobatus narinari*) leaped from the water and hit her. The collision knocked Zagorski backward onto the floor of the boat. She died from injuries to her skull and brain, not from the ray's poisonous barb.

"The force of that impact was dramatic," said Jorge Pino, spokesman for the Florida Fish and Wildlife Conservation Commission. Pino said collisions with humans are "unheard of. It was a freak, freak accident."

Uncommon, for sure. But stingray-related deaths do happen. They serve as a reminder not to be fearful of the ocean, but instead to remember that the creatures that live within it deserve our respect. As Steve Irwin once said, "We don't own the planet Earth, we belong to it. And we must share it with our wildlife."

"WE DON'T OWN THE PLANET EARTH, WE BELONG TO IT. AND WE MUST SHARE IT WITH OUR WILDLIFE."

—STEVE IRWIN, ZOOKEEPER, CONSERVATIONIST, AND TELEVISION HOST

CHAPTER 7

YOUR SHIP'S MADE OF WHAT?

BUILDINGS. PATIOS. STREETS. THESE ARE THINGS ONE MIGHT expect to be made of concrete.

But ships? Everybody knows that a concrete block thrown into a lake will sink straight to the bottom. It's ridiculous to think concrete can float! *Or is it?*

More than 175 years ago, a French man named Joséph-Louis Lambot began to make metal-and-wire support structures to be placed within his designs. He applied a layer of cement over the supports to build water tanks and feed troughs. Lambot's creations relied on the strengths of each product: Metal was extremely strong. Cement was affordable and could be easily shaped. This process of applying cement over wire-and-metal framework became known as ferro-cement. It was commonly used to create planters, roofs, sculptures, and more.

Lambot, however, was the first person to build rowboats using the process. They are thought to be the earliest concrete boats. The idea of a concrete creation actually floating was so unexpected that one of his boats was exhibited at the 1855 Paris Exposition.

Sound unbelievable? Grab a seat—it's time for a quick science lesson.

Buoyancy is a force that helps things float. When you put something in the water, like a boat or a beach ball, the object pushes some water out of the way. This is called displacement. The water doesn't like being pushed away, so it pushes back on the object. This push is the buoyant force. If the buoyant force (the water pushing up) is stronger than the weight of the object (the gravity pushing down), the object floats. If the object is too heavy and the buoyant force isn't strong enough, it sinks.

BUOYANCY IS A FORCE THAT HELPS THINGS FLOAT.

That explains why a boat made of something as heavy as steel—or even concrete—can float if it's the right shape and if it's displacing enough water. Lambot understood these principles, so building boats out of concrete wasn't a far-fetched idea to him.

Soon others started experimenting with ferro-cement boats. They built vessels of different sizes and shapes. In the 1890s, an Italian engineer named Carlo Gabellini built barges, which are long flat-bottomed boats, and small ships out of concrete.

In 1917, a Norwegian shipbuilder named Nicolay K. Fougner launched the first oceangoing concrete ship, an eighty-four-foot-long barge named *Namsenfjord*. Originally, government officials weren't so sure about Fougner's plans. He needed their approval if he wanted to build a seaworthy ship, one certified to carry a Norwegian crew and flag. So a special committee was appointed to review his proposal.

Fougner wrote at the time: "Naval architects, concrete engineers, and sailors expressed their unreserved opinions, through press and otherwise, that the whole idea was impossible and that a seagoing concrete ship was doomed to disaster the first time she encountered rough water."

He was more than happy to prove the doubters wrong. Fougner built about thirty concrete ships in total, often writing or speaking about their advantages over steel vessels. Concrete ships, he liked to point out, were cheaper to build, quicker to repair, and fireproof.

Fougner's earliest work on ferro-cement ships came as the United States entered World War I. At the time, most naval ships were made of iron or steel. Unfortunately, those same metals were required to manufacture items soldiers desperately needed: helmets, weapons, ammunition, tanks, and more. What to do?

Great Britain and Germany were building concrete ships. Should the United States do the same? The US government hired Fougner to do some research. He gave the thumbs-up, so US President

Woodrow Wilson approved the Emergency Fleet program. The country would build twenty-four concrete ships for the war effort. The plan was to use the ships as steamers and oil tankers, rather than as battleships or destroyers.

But setting up shipyards and training workers for a whole new construction process takes time. When the war ended in November 1918, work had only begun on half the fleet, and none of the concrete ships had been completed. Eventually, twelve ships were completed and sold to private companies. Several were used as steamships, one became an oil tanker, another was turned into a restaurant. Some of the ships were converted into floating oil barges or fishing piers. Others were sunk as breakwaters, barriers that help to reduce the force of waves and protect beaches or harbors.

Think that was the end of concrete ships? Think again.

When World War II began, steel was once again in short supply. It was 1942, and the United States decided to give concrete ships another try.

A Philadelphia firm called McCloskey & Company was hired to build two dozen concrete ships. Even though their sides were six inches thick, thirty years of improvements in ferro-cement construction meant these new ships would be lighter and stronger. They also were built faster, with a new one being launched every month beginning in July 1943. This time, all twenty-four concrete ships were completed before the end of the war. The ships finally saw some action and were used for transport, mainly as steamers or oil tankers.

EVEN THOUGH **THEIR SIDES WERE SIX INCHES THICK**, THIRTY YEARS OF IMPROVEMENTS IN FERRO-CEMENT CONSTRUCTION MEANT THESE **NEW SHIPS WOULD BE LIGHTER AND STRONGER**.

In March 1944, US Merchant Marine Richard R. Powers received an offer he couldn't refuse. If he would join the crew of a ship on its maiden voyage across the Atlantic, from Baltimore to England, the United States government would reward him with a return trip to the states on the RMS *Queen Mary*. Once the grandest ocean liner in the world, the *Queen Mary* had been stripped of her beautiful sculptures and murals when World War II began. During wartime, the ship ferried troops to and from combat zones, but her reputation lived on.

Powers was so awestruck by his potential reward that he didn't give much thought to the first part of the offer—until he boarded the SS *Vitruvius*.

"I couldn't help notice the ship was different from anything I had been on before," he wrote. "She was built of solid concrete. I wondered if this thing would float and get us across the Atlantic."

The ship wasn't actually made of "solid" concrete, but her reinforced concrete hull gave her a rough gray appearance—much different from the smooth metallic exterior of the steel ships

Powers was used to. When Powers saw lumber being loaded onto the ship, though, he became confident that the vessel was sturdy. He was heading to England.

These were war times, but the SS *Vitruvius* made its Atlantic crossing without being attacked. The ship was damaged, however, when another ship in its convoy accidentally rammed into it.

"I will never forget the sight of the bosun [a senior deckhand] getting a sack of concrete to patch up the cracks," Powers wrote years later. When the concrete ship docked in Liverpool, England, curious locals stopped by to see it. The merchant marine recalled one gentleman using his walking stick to tap on the ship's hull to "make sure it was really concrete."

On June 1, 1944, the SS *Vitruvius* again ventured out to sea. This time the ship and its crew were to take part in D-Day, the start of the campaign to liberate Europe and defeat Germany. They were among nearly 7,000 ships and smaller watercraft from eight allied countries participating in the Battle of Normandy.

As bombs blasted, sirens wailed, and planes buzzed overhead, the SS *Vitruvius* played a very different wartime role. Its crew was shuttled to safety, and it was sunk—on purpose. Its wreckage, along with several other intentionally sunken ships, was used to create an artificial harbor and docks. Their upper decks formed breakwaters that calmed the waves and allowed smaller ships to safely deliver soldiers and war supplies to the shore.

As World War II ended, so did the production of concrete ships. Their heavy bulk meant it took more fuel to power them, and they were prone to damage from collisions. When steel was once again readily available, it resumed its role as the preferred ship-building material.

Small ferro-cement boats are still being made today, mostly by hobbyists who love a good do-it-yourself project.

AFTER TAKING PART IN D-DAY, SEVERAL INTENTIONALLY SUNKEN SHIPS WERE USED TO CREATE AN ARTIFICIAL HARBOR AND DOCKS.

As for Merchant Marine Powers, well, he never got his much-hoped-for trip home on the *Queen Mary*. He and four others volunteered to stay in Liverpool to replace injured crew members on other ships. He did, however, play a role in a monumental day in world history on one of the most unusual naval projects of all time.

CHAPTER 8

THE HIGH PRICE OF WHALE POO

SCAT. DUNG. DROPPINGS.

Animal poop goes by many names and comes in a variety of forms. Most of the foul-smelling stuff, though, isn't particularly valuable. One spectacular exception? Ambergris.

No matter how you pronounce it—AM-ber-gris, AM-ber-gree, or AM-ber-grees—this waxy substance that forms in a sperm whale's intestines can go for a pretty penny.

Well, specifically, the rare whale excrement—often referred to as "whale vomit," despite the fact that it comes from the other end—can be worth $28 per gram (that's the weight of a paper clip). A basketball-sized lump of the precious poop could be valued at around $100,000. Some of these odorous lumps can weigh 100 pounds (45 kg) or more, and depending on its quality and age, ambergris of that size could sell for upwards of $1.5 million.

Ambergris isn't beautiful. It often looks like a regular rock, ranging in color from white to copper to dark gray. As it floats in salt water and is exposed to air and sunlight, it becomes lighter in weight and color, with its waxy outer coating turning gray or white. It may float for years before currents carry it to a beach.

The smell of ambergris varies from piece to piece. The scent of fresh ambergris is pungent. Thankfully, it mellows over time.

Scientist Chris Kemp became fascinated by ambergris and set out to research it. He wrote a book called *Floating Gold: A Natural (and Unnatural) History of Ambergris*. He says the smell of aged ambergris is pleasant. "It's a little like incense," he reports. "Almost like the smell of an old church."

Kane Fleury, a curator at New Zealand's Tūhura Otago Museum, often works with conservationists to examine whale remains. For him, the aged ambergris's smell is "strong, yet not bad" and contains woody, damp notes: "Although it has come out of the butt of a whale, it is not overly poo smelling. But it does smell strongly of marine mammal."

When people are willing to spend so much money on something, there's generally a reason. And in the case of ambergris, there are several. For starters, it's a prized ingredient in some of the world's most expensive fragrances. You read that right. There's whale poop in pricey perfume.

Companies like Chanel, Guerlain, and Lanvin use ambergris as a fixative; it helps scents stick to the skin and last longer. Legend has it that Elizabeth I, queen of England and Ireland, perfumed her gloves with ambergris because it maintained its fragrance for years, even after repeated washings.

Because ambergris is so rare, you might wonder whether a human-made product could do the same thing? The answer is yes. In 1950, scientists developed a compound called ambroxide, which most perfume companies now use. But many noses, the nickname for professionals who design perfumes, say the human-made replacement isn't quite the same. Kemp compared it to a song by a cover band or a fake designer purse. "It's close, but it's simply not the real thing."

While ambergris is best known for its role in the fragrance industry, it has been used for many purposes throughout the years. It has flavored coffee and hot chocolate and was an ingredient in the world's earliest known recipe for ice cream. It was once thought to ward off the bubonic plague. It has been burned as incense and used as a spice. There are even rumors that King Charles II enjoyed eating ambergris with eggs.

POSSIBLE
USES FOR
AMBERGRIS

FRAGRANCES (GREAT FIXATIVE!)

COFFEE FLAVOR BOOSTER

~~BUBONIC PLAGUE BLOCKER~~

HOT CHOCOLATE ENHANCER

~~ICE CREAM INGREDIENT~~

INCENSE

SPICE (APPARENTLY DELICIOUS WITH EGGS?)

The lore surrounding ambergris is another factor in its appeal. Although it has been traded for more than 1,000 years, not much is known about this magical manure. Early theories suggested ambergris might be hardened sea-foam or the product of underwater volcanoes. In ancient China, it was referred to as "dragon's spittle." People suggested it was produced by seals, crocodiles, or large birds. Some sailors believed ambergris grew like mushrooms on the ocean floor. Others thought ambergris might be fish liver or petrified fruit. Its connection to sperm whales wasn't discovered until the mid-1800s. Finally, in 2020, DNA analysis was used for the first time to prove scientifically that ambergris does indeed come from sperm whales.

Author Herman Melville included an entire chapter about ambergris in his novel *Moby-Dick*. He wrote: "Who would think, then, that such fine ladies and gentlemen should regale themselves with an essence found in the inglorious bowels of a sick whale!"

IN ANCIENT CHINA, AMBERGRIS WAS REFERRED TO AS "DRAGON'S SPITTLE."

Sick whales? It's a theory, but it still hasn't been proven. Marine biologist Robert Clarke spent more than fifty years studying ambergris and the whales that produce it. In 2006, he published a paper summarizing all he had learned.

One part of the mystery the British biologist was able to confirm was that sperm whales eat a lot of squid. Depending upon its size, appetite, and what's available, a sperm whale may eat 350 to 700 squid each day. Squid beaks are made of chitin and are hard, like a parrot beaks. It's pretty much impossible to digest this part of the squid, so most whales do their best to vomit out those tough parts.

Some beak pieces, however, make it into the whale's digestive system. These pokey parts can get stuck in the intestines. Clarke believed these bits eventually create a blob that fecal matter collects around. The blockage grows until the rectum stretches beyond repair and the whale perishes. Clarke died in 2011, but his theory is still widely accepted, though researchers contend that whales *can* pass the masses without necessarily dying. What all researchers seem to agree on is that the formation of ambergris is closely associated with the whale's squid-heavy diets. Most researchers also think that not all sperm whales experience these digestive issues, so just 1 to 5 percent of them are producing this "floating gold."

This strange-but-probably-true phenomenon explains why ambergris is so rare. There are only about 300,000 sperm whales left in the world, which means just 3,000 to 15,000 of them are producing ambergris.

And, as if its legend wasn't already mysterious enough, it's

THERE ARE ONLY ABOUT **300,000 SPERM WHALES LEFT** IN THE WORLD, WHICH MEANS JUST 3,000 TO 15,000 OF THEM ARE PRODUCING AMBERGRIS.

illegal to trade or possess ambergris in India, Australia, and the United States. These countries have laws that prohibit the sale of whale products. It's part of an effort to protect an endangered species. Even though it might be possible that sperm whales don't have to die to produce ambergris, there are people who would likely kill one to try to harvest the valuable waste. After all, in the 1800s, tens of thousands of whales were killed each year for their oil and bones.

Even in places where it is legal to gather ambergris, collectors are reluctant to talk about it. Death threats have been issued to those who bring outsiders to beaches known as good hunting grounds. Ambergris is valuable, and they don't want others to find treasure in their secret spots. The market is flooded with fraudsters who make money buying and selling fake ambergris. Dealers who purchase the stuff say they also deal with hopeful folks who bring what they *think* is ambergris only to find out it's a rock, fossilized wood, a lump of fatty waste, or—worse yet—dog poop.

The opportunity to study ambergris—and the sperm whales that produce it—is also a bit elusive. As temperatures rise and oceans get warmer, sperm whale migration patterns are changing. These evolving migration routes will likely affect where and how often ambergris comes ashore.

Another environmental change? In the past, ambergris was always formed around squid beaks. These days, though, the substance has sometimes been found formed around plastic pollution that collects in a whale's gut. That only deepens the mystery. Who is eating the plastic: the squid or the whale?

Mystery, rarity, legality, turf wars. Ambergris really does have it all. One thing is certain, though: The story of this inexplicably expensive excrement is still unfolding.

CHAPTER 9
TAKING ON THE TIDE

STEVE HOUSER IS A SURFER. HE AND HIS SURFING BUDDIES have spent years studying waves, paddling out to waves, and riding waves.

"Let's say I know enough about the ocean to truly respect it," he said. "I know its power and I know it can kill you if you're not careful. I'm always on the lookout for people who've gotten themselves into bad situations."

A bad situation is precisely what Gabe McCabe found himself in over Labor Day weekend in 2023.

Like many New Jersey residents, McCabe and his family had traveled to the shore for a last taste of summer. It was a hot day, perfect for walking along the beach and wading in the water. McCabe's not a swimmer, so he was careful not to venture too far from shore.

But remember: The ocean is powerful. And volatile.

Without warning, McCabe, his wife, and others were sucked out to sea in a rip current. Suddenly in deep water, McCabe was panicking and flailing. He quickly wore himself out as he tried unsuccessfully to battle the current. Things were not looking good, but luckily, McCabe was about to get some help thanks to the quick thinking of a man he had never met.

"I'd been in the water with my kids," Houser said. "I sensed right away that things weren't right. The water was getting really rough. As a surfer, I can spot a rip current pretty easily. I know the conditions for them, and I knew this had the potential to be dangerous."

He got his young children back on the beach and then headed into the water, toward McCabe. One of Houser's surf sponsors had given him an oversized boogie board. It just so happened that he had been testing it out with his daughter that day. The board had handles, which made it great for kids—and for saving lives.

"I could see a group of six or seven people who needed help," said Houser, a marine turned special education teacher. "I started waving for the lifeguards to come help and I swam toward Gabe because he seemed to be suffering the most."

Because he'd been wearing a camera to get footage of his sponsor's boogie board, Houser captured the entire rescue on video. The recording shows a frightened McCabe clinging to the board as Houser pulls him back to shore. Twice the duo was overcome by waves, and each time Houser encouraged McCabe to hold on.

"He didn't have much time left," recalled Houser. "He was panicky and on the precipice of drowning. Rip currents can be scary, even for good swimmers. I'm just glad I was in the right place at the right time."

Rip currents used to be called riptides or undertow, but experts say those terms are now considered incorrect. Rip currents are not tides. They are channeled currents of water flowing away from the shore. They can exist during any part of the tidal cycle. Rip currents also don't pull people under. They run horizontally, carrying people away from the shore like a treadmill.

They can be found on most any beach with breaking waves. Even lakes that have tides, including North America's Great Lakes and Lake Waihola in New Zealand, can have rip currents. Rip current

RIP CURRENTS ARE CHANNELED CURRENTS OF WATER FLOWING AWAY FROM THE SHORE.

speeds vary. Sometimes they're slow and aren't all that dangerous. Other times they may flow away from the beach at a rate of 5 miles (8 km) per hour—that's faster than an Olympic swimmer can swim.

Rip currents often form at low spots on the beach or along jetties or piers and can vary in width from 10 to 100 feet (3 to 30 m). They can be difficult to see, especially in rough ocean conditions.

"Rip currents are the most dangerous things at most beaches," said Stephen Leatherman, a coastal ecologist who also goes by the nickname Dr. Beach. "A lot of people are deathly afraid of sharks, but rip currents kill far more people than sharks."

It's true. Worldwide statistics are hard to come by, but in the United States, just one person on average is killed each year by sharks, while more than 100 people drown, and far more struggle, in rip currents. Many lifeguards refer to rip currents as "drowning machines."

The day McCabe was rescued, all swimmers made it safely out of the rip current and back onto the beach. But not everyone is so lucky.

A Pennsylvania family of eight traveled to Florida in June 2024. They were swimming along Stuart Beach when Brian Warter, Erica Wishard, and two of their teenage children got caught in a rip current. The kids were able to break the current and get to safety. They tried to help their parents, but conditions became too dangerous. Both Warter and Wishard drowned.

Former WWE star Shad Gaspard and his son, Aryeh, were in a group of swimmers pulled into a rip current at California's Venice Beach in 2020. A lifeguard went

IN THE UNITED STATES, JUST ONE PERSON ON AVERAGE IS KILLED EACH YEAR BY SHARKS, WHILE **MORE THAN 100 PEOPLE DROWN IN RIP CURRENTS.**

out on a two-person rescue, but when he reached the swimmers, Gaspard told the lifeguard to save his son first. "In sixty seconds, the lifeguard went back out to get Shad," Kenichi Haskett, the section chief of the area's lifeguard division, told *The New York Times.* "He saw him and then a wave came and pushed him under." Gaspard did not resurface.

In 2008, Scott Napper had a diamond ring in his pocket with plans to propose to his girlfriend, Leafil Alforque. The couple was walking along Oregon's Neskowin Beach when a wave splashed their legs. The next thing Napper knew, Alforque became caught in a rip current. "She was about thirty feet away, getting swept away," Napper said. "That's the last I saw of her."

These stories are heartbreaking. Often, it's tourists unfamiliar with the danger of rip currents who get caught in them. But rip currents can affect strong swimmers, too. They may get caught off guard and panic in the moment. And numerous people have died trying to help others.

Experts urge beachgoers to swim near lifeguards and not to swim alone. Look for signs or flags warning about rip currents or the potential for them. Spotting a rip current can be difficult. It's often easier to identify them from above, from a dune or hill. Rip currents run in the places where waves aren't breaking. It may also be easier to spot them if there's sea-foam or sediment in the water being transported away from the beach.

For those who find themselves caught in a rip current, the best thing to do is stay calm. Swimming back to shore against the current will only wear you out. The best way to swim out of the rip is to swim parallel to shore, then follow breaking waves back to shore at an angle.

> IF YOU'RE CAUGHT IN A RIP CURRENT, **SWIM PARALLEL TO SHORE**, THEN FOLLOW BREAKING WAVES BACK TO SHORE AT AN ANGLE.

"I highly recommend people have a flotation device with them in the ocean. That could be a life jacket, a surfboard, or a boogie board," advised Houser. "Even the best swimmers can get tired, and the ocean is unpredictable. Absolutely anyone can find themselves in a rip current, even if they know what they're doing. Being prepared can save your life."

CHAPTER 10
GRIEVING A LOSS

AN ORCA KNOWN AS J35 GAVE BIRTH ON JULY 24, 2018. HER calf lived less than an hour. It's sad, but it happens. A lot.

In fact, in 2017 researchers from the University of Washington found that only three out of every ten southern resident orca pregnancies result in a healthy baby.

The southern residents are the smallest of the resident orca populations. Resident orcas live in large pods and stay within their family groups their entire lives. They eat fish, as opposed to Bigg's orcas, which eat marine mammals. The southern resident orcas include the J, K, and L pods—seventy to seventy-five members in total. These orcas, with their distinctive black-and-white coloring, are members of the oceanic dolphin family. That's right: The creatures known as the southern resident killer whales or toothed whales are the largest members of the dolphin family. Females can weigh up to 8,000 pounds (3,600 kg), and males weigh up to 10,000 pounds (4,500 kg). They may live thirty to 100 years.

Perhaps the most important fact about southern resident orcas, though, is that they are an endangered species. When J35's baby died in 2018, her orca community had not had a surviving calf in three years.

The losses take an emotional toll on orcas and those who love them—and J35, also known as Tahlequah, was no exception. The twenty-year-old whale experienced a sadness so deep that the world grieved with her.

When Tahlequah's calf died, she kept it afloat for seventeen days. She balanced the baby's 400-pound (180-kg) body on her head or gently gripped it in her teeth as she followed her pod. They swam continuously, day after day in the waters between Vancouver, Canada, and the San Juan Islands in northwest Washington State. This 1,000-mile (1,600-km) journey became known as Tahlequah's grief tour.

Over the course of the two and a half weeks, the calf's body would occasionally dip below the water's surface. Each time, the weary mother orca would dive down and pick it up to prevent it from sinking. Over and over again. For days on end. When it appeared Tahlequah's own health was at risk because she wasn't eating, other orcas from her pod stepped in to help keep the calf's body afloat.

In the early days of Tahlequah's mourning, scientists thought this behavior was normal. Orcas are matrilineal—they spend almost their entire lives with their mothers. The bond is tight, perhaps because their pregnancies last eighteen months (twice that of humans). Prior to the death of Tahlequah's calf, orca mothers had been seen propping their dead calves on their foreheads in an apparent attempt to keep them with the pod. They'd been known to nudge their babies' bodies along for a few days or even a week.

But Tahlequah's devotion to her baby took things to a whole new level. And people began to notice.

The longer she carried her calf, the further Tahlequah's story spread. Local reporters in Canada and along the United States' West Coast were the first to write about Tahlequah's grief tour. Soon news anchors on nationwide programs such as *Good Morning America* and *Today* were providing daily updates. It wasn't long before Tahlequah was making headlines around the world:

"Not letting go: Grieving orca carries dead calf for days"—SBS (Australia)

"Grieving orca mother carries dead calf for days as killer whales fight for survival"—*The Guardian* (UK)

"Heart-wrenching sight: Grieving orca mother carries dead calf for days"—*The Daily News* (Egypt)

Many people were drawn to Tahlequah—orca lovers, animal lovers, people suffering their own losses. The *Seattle Times* newspaper asked readers to tell them how Tahlequah's story affected them. More than 1,000 people responded, some saying they were losing sleep thinking about the orca. A few readers wrote poems; one man wrote a song. People of all ages, from all over the world, related to Tahlequah's ordeal.

Conservationists and whale experts followed Tahlequah's story closely for another reason: They were concerned for the health of the grieving orca.

Scientists didn't want to get too close in case they further disturbed Tahlequah. That meant they couldn't tell if other whales were foraging for her and feeding her. An adult whale of Tahlequah's size can usually go without eating for a month. But no one was certain what her condition was before she gave birth. Was she underweight? Was she sick? Did she have an infection? What was her mental and emotional state?

Deborah Giles was working as the science director for a group called Wild Orca during Tahlequah's famous swim. Each time the baby's body fell beneath the surface of the water, Giles became more concerned. "What is beyond grief? I don't even know what the word for that is, but that is where she is," Giles told *The Seattle Times*. "She has to prime herself six, seven breaths to take a deep, long dive to go get that carcass. What is killing me is, when is it going to be the last time? And she has to make the decision not to go get it."

Two weeks passed. Finally, a plan was hatched to get food to the grieving orca. Members of the Lummi Nation, the original inhabitants of Washington's northernmost coast and southern British Columbia, teamed up with a local fish hatchery. They were able to load live Chinook salmon into a boat and take them to Tahlequah. Southern resident orcas' diets rely heavily on Chinook salmon.

"The mission was focused on feeding her, but it was also ceremonial," said Jay Julius, a member of the Lummi Nation. He explained that the Lummi have a centuries-old relationship with the orcas. "It was about honoring her and our other relatives who live under the sea. They were here before us. They teach us how to respect and navigate these waters."

Several observers wondered if orcas could truly feel human emotions like grief or loss. But Julius, who works full-time as a fisherman, said it was obvious to anyone who saw Tahlequah that she was grieving. When Tahlequah finally dropped her calf on August 9, 2018, it was as if a collective sigh could be heard from those riveted by her story.

> "WHAT IS BEYOND GRIEF? I DON'T EVEN KNOW WHAT THE WORD FOR THAT IS, BUT THAT IS WHERE SHE IS."
>
> —DEBORAH GILES,
> SCIENCE DIRECTOR OF WILD ORCA

"We were relieved for Tahlequah," said Julius. "But it wasn't like she was the only orca suffering. They are all vulnerable. We must be open to that truth."

And the reality was that this would not be the only time Tahlequah endured such a loss. Around Christmas 2024, scientists confirmed Tahlequah had given birth again, but by the new year, the calf had died. Once more, Tahlequah swam for weeks carrying her lost baby.

No one knows why Tahlequah's grieving was so prolonged. Some experts suggest that the short time the calves lived made her more deeply attached than if the calves had been born dead. No matter the reason behind her sorrow, Tahlequah's story drew public attention to a species at risk.

Julius and others hope that attention will translate into change and action. Orcas need salmon restored in the rivers that once provided them with a reliable source of food. They need clean, unpolluted water. They need protection from noise and boat traffic so they can rest, raise their young, and find food.

As for Tahlequah, her story isn't all sad. After each of her grieving periods, she seemed to rejoin J Pod as a healthy, active, and socially well-adjusted orca. And Tahlequah has two healthy offspring, including Phoenix, a whale born in September 2020, just two years after her first loss that gripped the nation's attention.

Monika Wieland Shields is president of the Orca Behavior Institute, based in Washington State. She said Tahlequah and Phoenix have a close bond, and while some other recent J Pod calves have been small, Phoenix is a good size for his age.

"He spends a lot of time with his older brother, J47 Notch, and he plays with J58 Crescent, who is the same age as him," Shields said, noting that within the orca community, Phoenix is known for his playfulness. The Orca Behavior Institute has posted videos of Phoenix balancing driftwood on his head, playing with kelp, squirting water out of his mouth, and breaching.

"Every healthy southern resident calf is a beacon of hope for the future, and Phoenix is no exception," said Shields, who adds that Tahlequah is also a source of optimism.

"We humans also heavily grieve the losses in the orca community," she said. "But Tahlequah was able to go through the tragedy she did and come out the other side, moving on to a brighter future for herself and her family. We, too, should learn how to mourn the losses and celebrate the joys in equal measure, while continuing to work hard to ensure that there's abundant salmon so future generations of southern resident killer whales can thrive."

CHAPTER 11
WHITE SHARK CAFÉ

GREAT WHITE SHARKS HAVE A BIT OF A REPUTATION, ESPECIALLY when it comes to appetite, bite, and fright.

- **THEY'RE BIG EATERS.** A full-grown great white eats more than 48 pounds (21 kg) of food each day—or just over 8 tons (8,200 kg) per year. And they're not picky. They'll eat fish, birds, seals, sea lions, and small whales.

- **THOSE TEETH ARE LEGIT.** Adult great white sharks have about 300 jagged teeth in as many as seven rows. The triangular teeth are sharp and can be more than 6 inches (15 cm) long.

- **SCARY? YES!** Though the movies make them look worse than they are, great whites are responsible for five to ten attacks on humans each year. But researchers want folks to know these encounters are because they're curious or confused—not because they actually want to chow down on people.

Still, there's a whole lot we *don't* know about these iconic predators, starting with: Why do they leave North America's West Coast for months at a time? How do so many of them meet in the same remote spot in the middle of the Pacific Ocean at the same time every year? And how in the world do researchers even know this is happening?

These massive sharks—which can reach lengths of nearly 21 feet (6.4 m)—are scientifically known as *Carcharodon carcharias*; they're also called white sharks, white pointers, or simply great whites. Scientists and surfers who were regulars in California's waters had long ago taken notice that the great whites that typically feed along the coastline there were disappearing every winter and reappearing in late spring.

Interesting, thought Stanford University marine biologist Barbara Block. Curious about this mass migration, she and a team of researchers started using electronic tags to track the great whites' movements beginning in the early 2000s. They tagged some sharks so often that they gave them names: Eugene, Leona, Tilden, Tom Johnson.

Block and researchers from the Monterey Bay Aquarium discovered that a large number of sharks

WHITE SHARKS, WHITE POINTERS, OR SIMPLY GREAT WHITES (**CARCHARODON CARCHARIAS**) ARE MASSIVE SHARKS THAT CAN REACH LENGTHS OF NEARLY 21 FEET (6.4 M).

that feast each summer and fall off the coasts of Mexico and California were leaving each winter. These sharks were swimming for up to 100 days to gather at a location roughly halfway between Baja California and Hawaii.

This open patch of ocean didn't seem to have any special qualities. Still, here were hundreds of sharks, gathered together in a space about the size of Colorado. It didn't seem to matter if they'd come from the southern tip of the Baja California peninsula or the chilly waters of San Francisco Bay; large numbers of sharks were meeting up in the same place at approximately the same time. Even stranger? When the males arrived at this secluded meet-up spot, they started diving up to 150 times each day, at all hours of the day and night. These weren't little dives; some of these sharks were diving to nearly 1,500 feet (450 m)—that's deeper than the Empire State Building is tall! Females dove, too, but not as often and mostly during daylight hours.

Researchers started calling this gathering spot the White Shark Café, although no one was really sure the sharks were eating there. In fact, satellite images of the area showed it had little food.

With few snacks in sight, new theories arose. Could sharks simply be showing off their diving skills? Might it be a mating area? Could it be a shark maternity ward, where mama sharks could give birth to bouncing baby sharks?

Scientists love a good mystery, but they love answers even more. So, in 2017, a team of scientists from five universities and research facilities was assembled; the initiative was again led by biologist Block. Her team attached trackable tags to the dorsal fins of thirty-seven great whites off the coast of California. A few months later, that same research team sailed to the café aboard a research vessel called *Falkor*. Their monthlong mission? To locate the tagged sharks and study their environment. They wanted to know why this seemingly unremarkable area was so important to the sharks.

The sharks' tags were programmed to pop off the sharks' fins and float to the surface when the *Falkor* arrived—and they did. Crew members searched the waters and found the tiny tags, which recorded not only where the sharks had traveled from and how long their journeys took but also environmental data such as temperature, pressure, and light. This information allowed researchers to better understand how white sharks travel through water. Thanks to those tags, sophisticated acoustic instruments, current sensors, and an unmanned underwater research vehicle, the scientists gathered data. Lots and lots of data.

One of the first things the researchers learned was that this area of open ocean was, contrary to earlier thinking, a well-stocked seafood buffet. It was teeming with deep-sea fish, squid, and jellies. Swirling waters here were pulling nutrients close to the surface and setting the scene for increased food production far below. Sharks

were using warm water currents to follow their desired prey down, down, down—which also explains the diving.

How could these super-smart scientists, with their years of experience, not have realized this before their research trip? That's easy. All the nutrient-rich plant life and fish exist deep under the surface of the ocean—too deep for satellites to detect. Plus, this was a part of the ocean that humans hadn't previously studied. Much like outer space, we are still developing the technologies that can explore the vastness and intense conditions that exist in the deep ocean.

In recent years, oceanographers, marine ecologists, and molecular biologists have joined the effort to learn more about White Shark Café, also known as the Shared Offshore Foraging Area (SOFA). They've got a new research vessel, *Falkor (too)*, a 363-foot-long (110-m) ship that's home to eight laboratories and state-of-the-art technology like remotely operated vehicles with high-quality video systems. They also conduct environmental DNA testing, which allows scientists to test DNA released into water to figure out if white sharks have recently swum by.

WHITE SHARK CAFÉ IS ALSO KNOWN AS THE SHARED OFFSHORE FORAGING AREA (SOFA).

Is it possible there are more spots like White Shark Café out there, undiscovered? Absolutely. Is there more to learn about the role this area plays in the life of great whites? Yes. And it's important information, says Block.

White sharks are apex predators. That means they're at the top of a food chain without natural predators of their own. Apex predators keep populations of other creatures balanced, and they control disease by eating sick or injured animals. Great whites also help spread important nutrients in the ocean. When they eat their prey, they break down the bones and hard parts, such as animal cartilage, releasing nutrients back into the water. These nutrients are used by tiny ocean plants and animals, which are the first links in the ocean's food chain.

Block believes that knowing more about white shark populations, life cycles, and migratory patterns is necessary to keep ocean ecosystems balanced and healthy.

As research continues, one of the primary goals of the *Falkor (too)* team is to protect the café.

> APEX PREDATORS (LIKE GREAT WHITES) **KEEP POPULATIONS** OF OTHER CREATURES BALANCED, AND THEY CONTROL DISEASE BY EATING SICK OR INJURED ANIMALS.

They've campaigned to have it identified as a potential World Heritage site. They'd also like to have it included in the list of marine protected areas (MPAs). These designations would restrict human activity and protect natural resources within the area. To receive these official protections, more information about the White Shark Café's biology and the sharks' behavior will be required, but Block is determined to do what it takes to safeguard the area.

Remember: Scientists love a good mystery. And now that they're confident that the White Shark Café holds so many answers, it's certain this feeding ground will continue to be studied for generations to come.

CHAPTER 12

NEW KIDS ON THE BLOCK

WHEN ERIN EASTON WAS A KID, SHE CAME ACROSS A BOOK about the *Titanic*, a real British ocean liner that hit an iceberg and sank in 1912. "I remember being fascinated by the blurbs about the animals they found living in the wreckage at the bottom of the ocean," she says.

Easton loved learning about creatures that thrive deep, deep below the ocean's surface: sea anemones, crabs, sea stars, and rattail fish. It was captivating, for sure. But she lived in Indiana, roughly 800 miles from the nearest ocean. Never in her landlocked childhood dreams did she imagine she'd one day be exploring the ocean's depths herself. But that's exactly what she does.

In 2024, Easton and Javier Sellanes led a team of scientists that explored seamounts along the coast of Chile. Seamounts are underwater mountains. Some even have volcanic activity. During this specific expedition, researchers discovered approximately 100 new species, including deep-sea corals, glass sponges, sea urchins, and squat lobsters.

How could a single research trip result in so many discoveries?

Well, our oceans are big—gargantuan, really. They are the largest living space on the planet and cover nearly 70 percent of Earth's surface. But the vast majority of them have never even been seen by humans.

As of 2024, about 240,000 marine species have been identified and named. That's a lot! But researchers think there could be another 2.5 *million* species out there to be discovered.

The waters Sellanes and Easton's team explored were largely uncharted. It would be new territory for everyone. Not so many years ago, scientists would head into areas like this and toss baited nets overboard in search of new sea creatures. There were obvious drawbacks to that process. For starters, the nets could only go down so far. Plus, many of the creatures they caught were crushed or lost as the nets were hauled back on board.

AS OF 2024, ABOUT 240,000 MARINE SPECIES HAVE BEEN IDENTIFIED AND NAMED. RESEARCHERS THINK **THERE COULD BE ANOTHER 2.5 *MILLION* SPECIES OUT THERE** TO BE DISCOVERED.

Researchers still use baited nets on occasion. More often, though, today's deep-sea exploration relies on technology. For example, there are autonomous underwater vehicles, or AUVs. Researchers design the path they want an AUV to follow. After being deployed, the AUV collects data that is stored inside the vehicle. When its preprogrammed mission is complete, the AUV rises to the surface, where researchers can retrieve it.

Sometimes scientists need to physically be in the water. But crushing water pressure and freezing temperatures make it impossible for scuba divers to explore deep-sea locations. Human-occupied vehicles, also known as HOVs, are a way around these problems. HOVs are submersibles that bring small groups of scientists and pilots, along with their electronic equipment, to the seafloor, allowing for in-person research and observation.

Researchers also use remotely operated vehicles, commonly referred to as ROVs. An ROV is connected to a research ship by a cable that provides power to the vehicle and sends real-time video back to the ship. The ROV's pilots stay on the ship and use controllers to send it to areas researchers want to explore. Many ROVs also have arms, or pincers, that can collect materials and

AUVs (autonomous underwater vehicles) are deployed on preprogrammed missons.

HOVs (human-occupied vehicles) bring small groups of scientists and pilots to the seafloor for in-person research and observation.

ROVs (remotely operated vehicles) are connected to a research ship by a cable that sends real-time video back to the ship.

water samples, and instruments that measure water clarity, light penetration, and temperature.

The Chilean expedition relied on an ROV called *SuBastian*. It was able to travel more than 2.5 miles (4 km) below the ocean's surface. *SuBastian*'s pilots directed the underwater robot to take video and collect samples of various plants and animals. It was an expedition that resulted in more discoveries than the crew could have hoped for.

"It felt like every time the ROV went around a rock outcrop there was something new to see," said Easton. "There was a lot of excitement. Team members kept yelling, 'Erin, Erin, look at this!' And every time we went around another rock outcrop, we'd see something we'd never seen before."

Sellanes agreed. "In general, it's quiet on the ship because you are with the pilots of the submersible and they need to concentrate," he said. "But when some extraordinary animals show up, everybody screams and cheers. You definitely hear lots of 'wows' and 'awws.'"

Full identification of these new species can take many years. Researchers must check scientific literature and collections in

> **MANY ROVS ALSO HAVE ARMS, OR PINCERS**, THAT CAN COLLECT MATERIALS AND WATER SAMPLES, AND INSTRUMENTS THAT MEASURE WATER CLARITY, LIGHT PENETRATION, AND TEMPERATURE.

museums. If they don't find any existing species that match the description of their discovery, they can propose a formal description of the new species—including a name. And for anyone out there who thinks, *Well, I'd name the cool new species after myself*, that's not allowed.

New species get two-part names. The first part is the genus—that's the part that tells scientists which species are related. The second part of the name identifies or describes the creature. Subspecies must have a three-part scientific name.

Of course, the expedition Sellanes and Easton led was just one of many deep-sea missions that have taken place in recent years. In fact, at nearly the same time that two separate expeditions were discovering new sea life near Chile, another crew was exploring the Bounty Trough, off the coast of the South Island of New Zealand. That team encountered another 100 new species. Their discoveries included dozens of mollusks, three fish, several species of coral, and a new type of shrimp.

THE FIRST NAME OF A SPECIES IS THE **GENUS**, WHICH TELLS SCIENTISTS WHICH SPECIES ARE RELATED. THE SECOND PART OF THE NAME IDENTIFIES OR DESCRIBES THE CREATURE. SUBSPECIES MUST HAVE A THREE-PART SCIENTIFIC NAME.

Ocean Census is helping to spearhead even more deep-sea exploration. The organization was set up in 2023. Its founders want to bring together some of the world's best researchers to speed up the exploration process. When Ocean Census began its work, the group set the ambitious goal of finding at least 100,000 new marine species within ten years.

Why do all these oceanic discoveries matter?

Simply put, oceans provide oxygen that humans need. Half the oxygen production on Earth comes from the ocean's plants, algae, and bacteria. These organisms use carbon dioxide, water, and energy from the sun to make food for themselves, releasing oxygen into the air as a by-product.

> HALF THE OXYGEN PRODUCTION ON EARTH COMES FROM THE OCEAN'S **PLANTS, ALGAE,** AND **BACTERIA.**

Oceans also play a major role in regulating Earth's climate. They absorb and store heat from the sun. Currents then redistribute heat and moisture, which can influence weather patterns. Understanding our oceans can lead to better hurricane, typhoon, cyclone, earthquake, and tsunami forecasting.

Scientists hope that finding new ocean species may also reveal new sources of food, energy, or even vaccines. They may even inspire inventions that mimic the ways deep-sea animals move, filter water, or grip objects.

"Expeditions like ours remind you how huge the oceans are and how much we still have to learn," said Easton. "It's my hope that we'll just continue building and building our knowledge. And, along the way, I hope these discoveries will convince people that we need to do more. We need to do whatever we can to protect and care for our oceans."

CHAPTER 13

THE (DOG'S) NOSE KNOWS

SCIENTISTS SAY A DOG'S SENSE OF SMELL MAY BE UP TO 100,000 times better than that of a human's. It makes sense. Dogs have about fifty times more scent receptors than people do. Plus, the part of a dog's brain that analyzes smell is about forty times larger than the comparable part of a human brain. That all adds up to making dogs super sniffers.

When trained, a dog can use its sense of smell to detect drugs or explosives, track down missing persons, point out mattresses in-fected with bed bugs, or sense low blood sugar in diabetes patients, among other skills. Even an untrained pup can tell from one sniff of a visitor's pant leg whether she has pets at home.

> DOGS HAVE ABOUT **FIFTY TIMES** MORE SCENT RECEPTORS THAN PEOPLE DO.

Eba, a mixed-breed rat terrier, is using her fine-tuned olfactory senses out on the high seas. You see, Eba is a conservation canine, a dog that works to sniff out scat from endangered orcas.

In other words, she's got a nose for orca poop.

Biologist Deborah Giles rescued Eba from the streets of Sacramento, California. Originally, Giles thought Eba would be like any other family dog. They'd go for walks, cuddle on the sofa, and play fetch.

Then, in 2019, the Conservation Canines program at the University of Washington lost funding for its scent-detection dog-trainer teams. This group had previously played a big role in tracking down orca poop so that researchers could learn more about these beautiful black-and-white beasts, but the program's dog and handler moved to another program. When the UW's funding was renewed, Giles, who works for the university, asked if Eba could try out her sniffing abilities. Fully aware that this dog might not have what it took, the university gave Giles and Eba the green light.

> EBA IS A **CONSERVATION CANINE,** A DOG THAT WORKS TO SNIFF OUT SCAT FROM ENDANGERED ORCAS.

Eba started training on land. First, she had to become familiar with the fishy smell of orca poop. Then she had to find the scat in increasingly tricky hiding places, all while learning to filter out other scents. Eba was a fast learner, mastering all that in just four days. Next came training on a boat. Giles hid scat in a bowl floating on the water. Eba couldn't run up to it, but she could run to the side of the boat the smell was coming from. She was a natural. Eba found her first whale scat just a day after the southern resident orcas returned to waters off the coast of Washington State.

"Not all dogs make good whale-scat dogs," Giles told the American Kennel Club. "Some get seasick or never get their sea legs, but Eba was perfect; she's been our main dog and I've been the main handler ever since."

While working, Eba wears a bright orange life jacket. When she detects the scent of orca scat, she uses cues like wagging her tail, twitching her whiskers, licking her lips, or sniffing with her nose in the air. Those signals direct both Giles and the boat captain toward the target.

> **WHEN EBA DETECTS** THE SCENT OF **ORCA SCAT, SHE USES CUES** LIKE WAGGING HER TAIL, TWITCHING HER WHISKERS, LICKING HER LIPS, OR SNIFFING WITH HER NOSE IN THE AIR.

Giles and Eba go in search of scat nearly every day from May to October, when the southern residents are off the coast of Washington. When they get a report of an orca sighting, they try to stay at least a quarter mile behind them. They don't want the noise from their boat to add to the orcas' stress. Giles says finding excrement from that distance without Eba's help "would be like looking for a needle in a haystack when the entire haystack is moving and so is the needle." Eba can locate orca excrement in the water up to one mile away.

Poo detection is a fast-paced treasure hunt. Orca poop may be red, white, green, brown, yellow, or tan in color. Adult male orcas can weigh more than 13,000 pounds (5,900 kg), but that doesn't mean their scat is easy to find. Giles says it may be as large as a dinner plate, or as small as a silver dollar. No matter the color or size, it only floats for a few minutes. If you're too slow, the poop will sink, taking with it valuable health information.

"If we're moving into the scent cone, she'll be right at the front of the boat, leaning over and, as we pass by the heartiest smell, she'll whip around to the side," said Giles. "And that's when we know to turn [to locate the excrement]."

ORCA POOP MAY BE RED, WHITE, GREEN, BROWN, YELLOW, OR TAN IN COLOR. IT MAY BE AS LARGE AS A **DINNER PLATE**, OR AS SMALL AS A **SILVER DOLLAR**.

Once poop is within reach, a scientist on board uses a bucket on a long pole to scoop it out of the water. It's placed in tubes and stored in a cooler until the boat docks. Occasionally Eba will find as many as ten samples in a day, but more often it's fewer. Some days the team doesn't find any. Eba knows not to bark at the orcas and not to jump in to swim with

them. A job well done earns her a game of tug—her favorite. Eba's unusual abilities have made her a bit of a celebrity. She's been featured in magazine and newspaper articles, and on PBS, the BBC, and *It's a Dog's Life* on Disney+.

Cool talent, but why is finding orca poop important?

Analysis of the excrement samples can provide detailed information about the animals, from their sex and stress levels to what they've been eating. Scientists can check the orcas' hormone levels to learn if they're pregnant. The samples may reveal information about the presence of parasites and pollutants. DNA can also be extracted from the scat. That allows scientists to track individual animals, which is important for endangered species like the southern resident orcas.

Studying whale poop may not sound super appealing, but scientists say it's far superior to other methods they've used to gather information. In the past, scientists have used crossbows, spears, or air guns to shoot darts

EXCREMENT ANALYSIS DATA SHOWS . . .

- SEX
- STRESS LEVELS
- DIET
- HORMONE LEVELS/PREGNANCY
- PARASITES
- POLLUTANTS
- DNA EXTRACTION FOR INDIVIDUAL TRACKING

at orcas. The darts had cutting tips that pulled small plugs of skin and blubber from the animals, which scientists could analyze. These sample-gathering methods are still used, but scat collection is less invasive. No one has to get too close to the whales, and no poking or prodding is required.

Eba's work is important, but she's not the first nor the only conservation canine.

Fargo was a Rottweiler trained to find feces from right whales along the eastern coast of the United States and Canada. Past orca scat sniffers have included Waylon, a yellow Lab; Tucker, a black Lab mix; Sadie, a flat-coated retriever; and Jack, an Australian cattle dog.

In addition to seagoing conservation canines, many other pups are putting their skills to work on dry land. Since the 1990s, researchers have trained dogs to track down scat belonging to animals ranging from wolves, bobcats, and coyotes to the minuscule droppings of New Mexico mountain salamanders and Pacific pocket mice.

Many of the best conservation canines are pups who are considered too hyper to make good pets. They become so focused on an activity like fetch that they ignore

RESEARCHERS HAVE TRAINED DOGS TO TRACK DOWN SCAT BELONGING TO **WOLVES, BOBCATS, COYOTES, NEW MEXICO MOUNTAIN SALAMANDERS,** AND **PACIFIC POCKET MICE.**

food or treats. Families grow frustrated, and these dogs often end up in shelters. But that focus is precisely what has helped Eba and others become valued wildlife researchers. Intensity and single-mindedness are their superpowers as they take on new missions: helping scientists save other species.

CHAPTER 14
STRANDED AT SEA

STEVEN CALLAHAN REMEMBERS A HURRICANE PUMMELING THE
Massachusetts coast when he was just a kid. Power lines snapped, trees swayed and bent like blades of grass, and the family's once-sturdy tree house was torn to bits.

Witnessing nature's fury impacted the young boy. He began tucking some cash, a jackknife, and a fishing reel into a box, and hiding it in a desk drawer. It was his first emergency kit.

"If disaster struck, I would be ready," recalled Callahan years later. "If anyone survived, it would be me."

It was as if young Steven Callahan could see into the future. It was as if he knew he was going to one day face the kind of disaster most folks would not—could not—endure.

Plenty of people own boats and enjoy an occasional weekend on the water. Callahan is not that kind of sailor. Rather, he has eagerly greeted rough waters and new coastlines for as long as he can remember. By the time he turned thirteen, he was dreaming of designing and building his own boats. In high school, he started sailing solo, helped construct a forty-foot trimaran, and taught himself to navigate using the sun, moon, stars, and planets as positioning points. After college, he was able to achieve his childhood goal as he repaired, designed, and built boats. He even taught boat design and illustrated textbooks on the subject.

In 1980, Callahan built a cruiser he called *Napoleon Solo*. Years later, he wrote a book called *Adrift,* in which he gushed about his creation: "*Solo* became much more than a boat to me. I knew her every nail and screw, every grain of wood. It was as if I'd created a living being."

> "*SOLO* BECAME MUCH MORE THAN A BOAT TO ME. I KNEW HER EVERY NAIL AND SCREW, EVERY GRAIN OF WOOD. IT WAS AS IF I'D CREATED A LIVING BEING."
>
> —STEVEN CALLAHAN, SAILOR AND AUTHOR

It was in that boat that Callahan set out to fulfill his dream of a cross-Atlantic voyage. In spring 1981, he loaded everything he owned onto *Solo* and began to sail from Newport, Rhode Island, to Bermuda, an island nation about 600 miles off the coast of North Carolina. After safely arriving in Bermuda, a pal joined Steven, and they set sail for England. The weeks-long voyage was successful. "Exhilarating," proclaimed Callahan.

After a bit of rest and after his buddy went home, Callahan entered the Mini Transat Race. The solo sailing competition would take sailors from Penzance, where the English Channel meets the Atlantic Ocean, to the Canary Islands, off the coast of north-western Africa, and finally to the white sandy beaches of Antigua, where the Atlantic Ocean meets the Caribbean Sea.

THE **MINI TRANSAT RACE** WAS A SOLO SAILING COMPETITION THAT TOOK SAILORS TO **PENZANCE**, THE **CANARY ISLANDS**, AND THE WHITE SANDY BEACHES OF **ANTIGUA**.

Initially, sailing was smooth. Soon, though, winds picked up and *Solo* was forced to leap over ten-foot waves. Rough seas and floating debris cracked the boat's hull. Callahan made quick repairs, but the race was over for him. He headed to La Coruna, Spain. He was not alone; at least seven other boats from the race were there.

Repairs were made and supplies were gathered. On January 29, 1982, Callahan and *Solo* were ready to sail again. No race this time—just a man and his boat headed toward the Caribbean.

Early days of the voyage went well. Seas were calm. Callahan rested on the deck, the gentle breeze rustling the pages of his book. If the weather held, he would be at his destination by February 25. Too bad Mother Nature had different plans.

Seven days into his trip, strong winds began to blow. Angry waves slapped at the boat's deck. *BANG!* Callahan crashed into something. Later, he speculated it may have been a whale or large shark, but in the moment, there was no time for guessing. It was the middle of the night, and Callahan jumped from his bunk. Water was pouring into his below-deck cabin. What happened? Where was the water coming from? The nose of the boat tipped downward. Oh no. *Solo* was sinking.

Callahan's mind raced, and his heart pounded loudly. He tossed his emergency raft overboard, inflated it, and dove in. But as he watched *Solo* begin to sink, he knew he had to return to the boat to gather supplies. Aching and cold, he climbed back on board. He grabbed an emergency duffel bag, water, flares, a spear gun, and a wet sleeping bag. Random items floated through the cabin, and he gathered what he could: eggs, an empty coffee can, a cabbage. He shuddered at the thought of being stranded on this tiny raft in the vast ocean for a few days or a week. Little did he know, the situation was more dire than he could imagine.

Over the next ten weeks—yes, *weeks*, not days—Callahan pushed his survival skills to their extremes. He stayed alive by spearing fish. He ate their meat raw, sucking nutrients from their organs and liquid from their eyeballs. He built a makeshift still, or a device that pulls the salt out of ocean water, providing him with about two cups of drinkable water every day. Waves splashed

into the raft, so he was forced to spend hours each day bailing water out. When his raft developed a four-inch hole, the former Boy Scout patched it by stuffing it with sponge and sleeping bag lining and tying it off with rope. Stretching and yoga provided relief for the sores he had developed from sitting and sleeping on wet rubber.

A **STILL** IS A DEVICE THAT PULLS THE SALT OUT OF OCEAN WATER.

Callahan kept a log in tiny notebooks that he kept dry in plastic bags. He recorded the weather, his mood, and battles with violent storms and curious sharks. He wrote lovingly about the fish he saw along the way, calling them "doggies." He was always hungry, always tired, always making repairs to his fishing gear and raft. Callahan shot off flares to signal passing ships, but time after time they went unnoticed. His thoughts ran dark. Hope was fading.

Finally, on day seventy-six, Callahan was spotted by three fishermen who spoke no English. They took him to a small Caribbean island called Marie-Galante. He was dehydrated and unsteady on his feet, and he had lost one-third of his body weight—forty-four pounds. But his story had a happy ending. He had faced a nearly insurmountable disaster, and he had lived.

Is Callahan the only person to survive being stranded at sea? No. Some other notable names who overcame similar dangers:

POON LIM survived 133 days on a wooden raft after his ship was hit by two torpedoes in 1942. Lim was the only person from his ship's fifty-five-member crew to survive. He was awarded the British Empire Medal in honor of his courage.

In September 1983, American sailor **TAMI OLDHAM ASHCRAFT** and her fiancé, **RICHARD SHARP**, began sailing from Tahiti to San Diego, California. A massive hurricane capsized their boat, and Sharp was lost in the storm. A badly injured Ashcraft was able to right her ship and rig up a crude sail. She spent forty-one days adrift in the Pacific Ocean, finally docking in Hilo Harbor on the island of Hawaii.

JOSÉ SALVADOR ALVARENGA, a fisherman from El Salvador, washed ashore on a nearly deserted island in the Pacific in January 2014. He had been stranded for 438 days and kept himself alive by catching fish with his bare hands and swallowing jellyfish whole.

There are others, but this survivors club is small. The ocean can be fierce, and the chances of lasting more than a few hours without supplies and shelter are slim. The United States Coast Guard says most rescues occur within the first thirty-six hours. According to the organization, without signaling devices "there is almost no chance of being spotted and ultimately rescued." In Callahan's case, the Coast Guard never really searched for him. Officials said they couldn't be certain he was really missing. Then, after pieces of *Solo* were discovered, they assumed it was too late, and there was no chance he was still alive.

Living through such a traumatic event tends to change a person. Callahan says it certainly changed him. Yes, he still sails, and he still eats fish, but his perspective has shifted.

THE UNITED STATES COAST GUARD SAYS **MOST RESCUES OCCUR WITHIN THE FIRST THIRTY-SIX HOURS.** WITHOUT SIGNALING DEVICES, "THERE IS ALMOST NO CHANCE OF BEING SPOTTED AND ULTIMATELY RESCUED."

"Whenever I fill a tub with water, I think, *This is more fresh water than what I lived on for two and a half months*," he wrote in *Adrift.* When someone exclaims, "I'm starving," he can't help but think, *Well, not really.*

In an interview with the global news organization *The Guardian*,

Callahan said he does not regret his seventy-six days alone at sea. "To this day I feel enlightened by what I went through because it changed me for the better," he said. "But would I want to be adrift in the ocean again? No way."

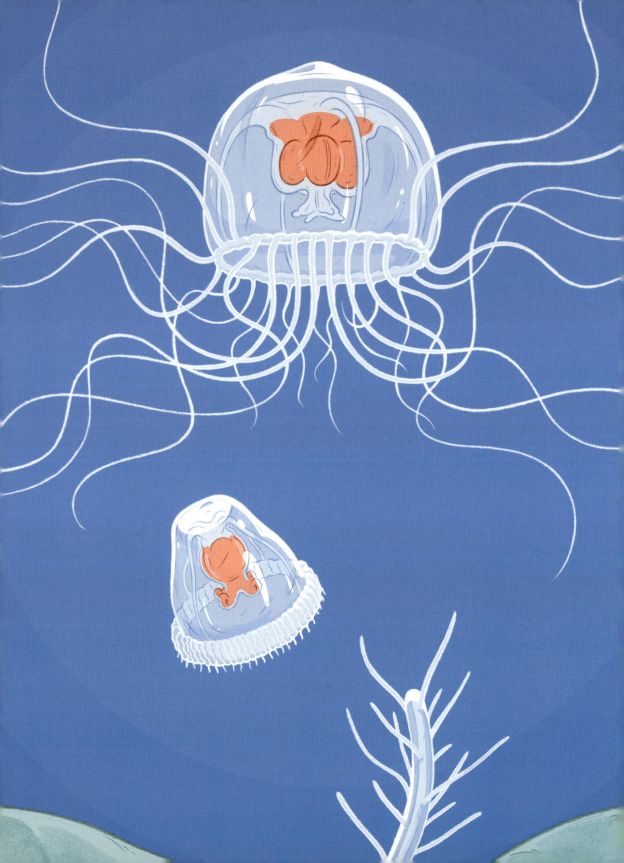

CHAPTER 15
OCEAN ELDERS

JEANNE CALMENT LIVED A LONG LIFE. A VERY, VERY LONG LIFE.
When she died in 1997, she was 122 years, five months, and fourteen days old—that's longer than any other human has lived to this point. The secret to her longevity? She gave credit to exercise, a diet rich in olive oil, and laughter.

But Calment was only a child compared to the Greenland shark.

Scientifically known as *Somniosus microcephalus*, these slow-moving sharks have an average lifespan of 250 years. Researchers believe some may reach more than 500 years old, making them the world's longest-living vertebrates. That's old! So old that some of the Greenland sharks swimming in the

> GREENLAND SHARKS (*SOMNIOSUS MICROCEPHALUS*) ARE SLOW-MOVING SHARKS THAT HAVE AN AVERAGE LIFESPAN OF 250 YEARS.

ocean today were alive in 1597, when Shakespeare wrote *Romeo and Juliet*. It's likely that some of today's living-and-breathing sharks were alive in 1664, when Isaac Newton began conducting experiments with gravity, and in 1793, when workers first began building the United States Capitol.

Greenland sharks would have seen a lot—if they could see. There are a couple of reasons why the sharks might be (metaphorically) squinting. For starters, they can live in waters up to 7,200 feet (2,200 m) deep. These depths are called the hadal, or hadopelagic, zone of the ocean, an extreme environment with low temperatures and pitch-black darkness. Greenland sharks can see when they're born, but small parasites firmly attach themselves on to their eyes, fluttering like wormy eyelashes. Researchers believe the parasites produce a substance that dissolves the sharks' eye tissue. Thanks to these clingy creatures, most Greenland sharks are blind or nearly blind by the time they reach adulthood. Like most deep-sea organisms, Greenland sharks have a strong sense of smell—a good thing, because otherwise they wouldn't be able to find food or figure out where they were going.

GREENLAND SHARKS CAN LIVE IN WATERS UP TO 7,200 FEET (2,200 M) DEEP. THESE DEPTHS ARE CALLED THE **HADAL**, OR **HADOPELAGIC**, ZONE, AN EXTREME ENVIRONMENT WITH **LOW TEMPERATURES** AND **PITCH-BLACK DARKNESS**.

Speaking of adulthood, Greenland sharks are not quick to mature. These deep-sea senior citizens grow less than half an inch (1 cm) per year. Female Greenland sharks don't start having babies until they are about 150 years old. When they do give birth, they often have litters of up to ten pups.

Despite its name, the Greenland shark is not only found in the waters near Greenland. The species has been reported as far east as France and Portugal and as far south as North Carolina.

For some shark species, like the great white or whale shark, scientists can count the growth bands on their vertebrae to determine their age. It's a process a lot like counting the rings on a tree. This exact method doesn't work for the Greenland shark, which is mostly made of cartilage and doesn't have hard body parts where growth layers can be deposited. So, instead, researchers use a scientific method called carbon dating to determine age. The Greenland shark's eyes contain proteins that are formed before birth. These proteins are preserved and don't decay or disintegrate as the sharks get older. Scientists can determine the age of a Greenland shark by carbon-dating these proteins.

> FOR SOME SHARK SPECIES, LIKE THE GREAT WHITE OR WHALE SHARK, SCIENTISTS CAN COUNT THE **GROWTH BANDS** ON THEIR VERTEBRAE **TO DETERMINE THEIR AGE.**

Scientists have spent years trying to figure out how and why these unusual creatures live so long. Many thought it was because they live in frigid water—as cold as 29 degrees Fahrenheit (−1.8 degrees Celsius). But new research shows it may have more to do with metabolism. That's the process that turns nutrients into energy and allows that energy to build and repair tissues. In most animals, humans included, metabolism slows down with age. Slower metabolism is why grandparents typically have less energy, less hair, and move slower than their much younger grandchildren.

Scientists recently discovered that Greenland sharks' metabolism never changes. Because of that, their ability to move, react, and recover from injury doesn't lessen as they get older. It's the same whether they're four or 400 years old.

Of course, Greenland sharks aren't the only oldies but goodies in our oceans.

The ocean quahog (*Arctica islandica*), for example, is an edible clam that lives in the Atlantic Ocean. This specific type of mollusk can live more than 400 years. Once the ocean quahog reaches maturity at around eight years old, it stops showing signs of aging.

THE OCEAN QUAHOG (ARCTICA ISLANDICA), IS A CLAM THAT CAN LIVE MORE THAN 400 YEARS.

In 2006, scientists were doing research off the coast of Iceland. They were analyzing the lines on clam shells to estimate the age of various clams. That's when they found Ming, a quahog clam they said was 402 years old. They called it Ming after the Chinese dynasty during which it was born.

Unfortunately, to determine the age of a clam, a section of its shell must be taken and studied. When the clam is opened, it dies. Ming's death triggered headlines that labeled the whole event "scandalous" and called the scientists "murderers." The scientists defended themselves, explaining that Ming was one of many clams being studied. It just happened to be very old.

The fury over Ming's death had died down by 2013. That's when scientists reexamined the shell using more precise techniques. It's now believed Ming was 507 years old when it died.

Bowhead whales (*Balaena mysticetus*) are one of the few whale species that live in chilly Arctic waters all year long. The bowhead is well suited to the cold; they have an insulating layer of blubber that can be up to 18 inches (46 cm) thick. Bowheads have very large heads and thick bodies. The bowhead's skull allows them to break through thick sea ice. They've even been seen surfacing through two feet of ice.

Bowheads are among the heaviest animals on Earth, weighing up to 100 tons (90 metric tons). For comparison, one bowhead whale weighs about the same as twenty-two male Asian elephants.

BOWHEAD WHALES (*BALAENA MYSTICETUS*) LIVE IN CHILLY ARCTIC WATERS ALL YEAR LONG. IT IS BELIEVED THEY CAN LIVE MORE THAN 200 YEARS.

Researchers believe these beasts can live more than 200 years, nearly twice as long as the oldest Asian elephant.

Immortal jellyfish (*Turritopsis dohrnii*) don't really fit neatly into an "oldest" or "longest lifespan" category. Rather, these nearly transparent jellies may be the "most everlasting" creatures in our oceans.

Immortal jellies are tiny—roughly the size of a Tic Tac mint. They were first discovered in the 1880s in the Mediterranean Sea. Scientists believe they are the only uniquely enduring creatures in our oceans.

THE IMMORTAL JELLYFISH (*TURRITOPSIS DOHRNII*) USES A RARE PROCESS CALLED TRANSDIFFERENTIATION, WHICH ALLOWS IT TO LEAP BACKWARD IN ITS DEVELOPMENT PROCESS INDEFINITELY.

Thanks to a rare process called transdifferentiation, the immortal jelly can leap back in its development process. Threatened by a predator? Stressed? Starving? The immortal jelly hits the reset button and reverts to its younger polyp form. It basically starts its life cycle over again. The new polyp buds and becomes a bell-shaped, tentacled creature that's genetically identical to the adult it used to be.

Knowing about immortal jellies, 500-year-old clams, and other long-living ocean creatures is amazing. But learning about them could have value beyond simply collecting odd facts.

Acknowledging the history of our oceans can help protect them moving forward. By understanding the changes they've endured over time, we are better able to identify threats, such as pollution, overfishing, and global warming. It also allows us to establish more conservation-minded policies. Additionally, studying these oceanic elders could someday help scientists better understand human metabolism. Or it might help doctors figure out how to cure a disease or replace cells that have been injured or damaged.

It's a win-win. And the longer we research these creatures, the more we learn. Because after all, with great age comes great wisdom.

CHAPTER 16
RESCUE BY TAXI

THANKS TO A KINDHEARTED TAXI DRIVER, ENDANGERED seabirds in New Zealand have been getting a lift to safety right when they need it most.

Hutton's shearwaters (*Puffinus huttoni*) are medium-sized brown-and-white seabirds with dark pink webbed feet. They spend about 90 percent of their lives at sea and are the only seabird in the world that raises its young in the mountains. Hutton's shearwaters make their nests high in New Zealand's Kaikōura Ranges and then migrate to the warm coastal waters of Australia. When they are five to six years old, they return to the Kaikōura Ranges to breed.

> HUTTON'S SHEARWATERS (***PUFFINUS HUTTONI***) ARE THE ONLY SEABIRD IN THE WORLD THAT RAISES ITS YOUNG IN THE MOUNTAINS.

Scientists believe fewer than 500,000 Hutton's shearwaters are left in the world. In the 1960s, there were eight shearwater breeding colonies. Now there are only two. Researchers point out several key reasons for their declining population. New predators such as feral pigs have been introduced into the region. A 2016 earthquake caused landslides, which destroyed much of the birds' nesting area. And plastics in the ocean are sickening and killing the shearwaters.

> SCIENTISTS BELIEVE FEWER THAN 500,000 HUTTON'S SHEARWATERS ARE LEFT IN THE WORLD.

There's also the problem of light disorientation, which is where the taxicab comes in.

Baby Hutton's shearwaters are chubby and covered with gray fluff. The Māori, the Indigenous people of New Zealand, have a word for these chicks: tītī. The tītī are cared for by both parents. They spend their early days snuggling in feather-lined nests made of grass and sticks. The nests are tucked into burrows high in the mountains.

After about eighty days, the fledglings' fluff is replaced with feathers. Around this development stage, their parents stop delivering daily fish dinners. It's time for these young birds to leave the nest—literally. From the first moment they leave the colony, they are on their own. No calls to Mom and Dad for help, no supportive wing to lean on. Which is rough, because that first twelve-mile flight from the mountains to the ocean is a challenging one.

Hutton's shearwaters are nocturnal. That means they're most active at night. So it makes sense that their first flights almost always begin after dark. And that's when the disorientation comes into play.

Remember, these are birds that have been living in burrows, which are basically holes in the ground or the side of a cliff. They've spent most of their early life in the dark, only peeking out of the burrows when their parents arrived with food or to practice flight motions during the end of their nesting period. When they

> HUTTON'S SHEARWATERS ARE **NOCTURNAL.** THAT MEANS THEY'RE MOST ACTIVE AT NIGHT.

fly over the coastal town of Kaikōura that very first time, it makes sense that the bright lights might confuse or even temporarily blind the young birds. Are they flying too high? Too low? Are they already at the ocean? They really have no idea where they are until—*thump!*—they crash into something or simply fall, exhausted, to the ground.

Local cab driver Toni Painting to the rescue.

Painting first noticed the dazed and confused chicks in 2015. Realizing the birds were in distress, Painting found a solution. She and her crew of volunteers scour the streets of Kaikōura, ready to transport the disoriented birds to a nearby rehabilitation center. Workers there check the birds for injuries and warm them. They provide fluids and give the birds time to rest and regain their strength before taking them out to sea.

Experts believe the seabirds see lights reflecting off shiny highways and think it's the surface of the ocean. The crashes understandably daze and may injure the shearwaters. Even if a bird appears to be in good shape, flying isn't always an option. A Hutton's shearwater's takeoff relies on thermal updrafts—rising pockets of warm air—from a wave or hillside to take flight. They aren't built to take off from flat ground.

Without rescue, the birds are likely to be hit by cars, starve, or be eaten by dogs, cats, or wild animals. And so Painting and her team patrol each fledging season, throughout most of March and April, collecting the wayward birds.

"I go out half an hour after dark," she told the British newspaper *The Guardian.* "Then I go every hour until half past midnight. It takes them half an hour to get down from the mountains."

Painting keeps a box in her taxi to hold the chicks. Typically, ten to twenty birds are found each night on roads along the coast. On their

> "I GO OUT HALF AN HOUR AFTER DARK. THEN I GO EVERY HOUR UNTIL HALF PAST MIDNIGHT. IT TAKES THEM HALF AN HOUR TO GET DOWN FROM THE MOUNTAINS."
>
> —TONI PAINTING,
> LOCAL CAB DRIVER AND BIRD RESCUER

busiest night, her team worked until dawn and rescued more than 200 shearwaters.

"I just love birds and I know they're threatened, so it was on my heart to save the species," Painting said.

Hutton's shearwaters aren't the only birds troubled by bright lights. Far from it.

Short-tailed shearwaters are common seabirds. They nest around Australia in the summer and then migrate to Alaska's Aleutian Islands or Russia's Kamchatka Peninsula in the winter. That's an 8,000-mile journey. Along the way, bright human-made light sources can be a fatal attraction to the birds.

Wilson's storm petrel is another seabird that frequently finds itself blinded by lights. These small dark brown birds with white tails live in all the world's oceans except the North Pacific and Arctic. Petrels navigate using landmarks in the night sky. Scientists believe that's why artificial light confuses them—they can't tell a streetlight from the moon or the stars.

Not all grounded birds can count on being rescued by a wildlife-loving taxi driver, but rehabilitation programs are in place in areas where large numbers of seabirds are known to crash.

Many cities and states also have "lights out" campaigns to educate residents and minimize risk to birds. In the United States, Audubon's Lights Out Program is a national effort to address the problem. Where businesses aren't willing to dim their lights, environmental groups are sometimes forced to step in. That's what happened in Hawaii in 2024. Two groups teamed up to ask a judge to force a popular Maui resort to turn off its bright outdoor lights

> MANY CITIES AND STATES HAVE **"LIGHTS OUT"** CAMPAIGNS TO EDUCATE RESIDENTS AND MINIMIZE RISK TO BIRDS.

in October and November, when Hawaiian petrels make their first flights from their nests. That battle is still working its way through the courts.

As for Toni Painting, she plans to keep rescuing as long as the seabirds need it.

"They're lovely birds," she said. "They're easy to pick up and release. And it just gives me a lot of pleasure, releasing the birds into the sea, where they are supposed to be."

CHAPTER 17
ERUPTIONS

YOU CAN HEAR THE PASSION IN KENNA RUBIN'S VOICE WHEN she starts talking about volcanoes.

"They are so cool," she says. "Yes, they're dangerous and destructive, but eruptions are just so exciting. I don't know many people who aren't awed by a fiery, explosive mountain spewing red-hot lava."

Rubin pauses and gathers herself. She wants to give volcanoes like Mount Ruang, Volcán de Fuego, and Mount Saint Helens the respect they deserve. She appreciates the ways in which scientists are monitoring these and other volcanoes using satellites, thermal imaging, and real-time video. But she wants to make it very clear that her true passion these days has to do with the volcanoes people can't see—the ones underwater.

"It's a fact many people don't know," says Rubin, a professor of oceanography at the University of Rhode Island. "Most volcanic activity on the planet occurs in the oceans, not on land."

It's true.

Volcanoes that exist on Earth's surface are called terrestrial volcanoes. *Terrestrial* means living or growing on land. There are more than 1,500 active terrestrial volcanoes on Earth, and about sixty of them erupt each year. That may seem like a whole lot of erupting—until you compare it to what's happening underwater.

Submarine volcanoes are exactly what they sound like—volcanoes that are located under the ocean's surface. How many of them are there? A lot. Several thousand active submarine volcanoes have been discovered, and scientists think there could be more than a million of them.

The general public doesn't hear much about submarine volcanoes because most of their eruptions barely make a ripple on the ocean's surface. See, the weight of the water above a submarine volcano creates extremely high pressure. That pressure keeps the majority of submarine volcanoes from erupting upward. Instead, the volcanoes have what's known as passive lava flow, where the lava flows along the seafloor.

VOLCANOES THAT EXIST ON EARTH'S SURFACE ARE CALLED **TERRESTRIAL** VOLCANOES. **SUBMARINE** VOLCANOES ARE LOCATED UNDER THE OCEAN'S SURFACE.

As eruptions occur, the lava flow cools and can cause some of these underwater volcanoes to grow taller. The taller they are, the less water they have above them pushing downward. That means volcanic gases and steam can more easily reach the surface of the water above. If conditions are right, the volcanoes can even blast rock straight out of the ocean.

All this submarine volcanic activity over thousands to hundreds of thousands of years can add up, to the point where islands are formed. Hawaii, Indonesia, the Galápagos, and Iceland are examples of islands formed by submarine volcanoes. Niijima is a newer volcanic island. It was formed in 2023 in the Pacific Ocean, off the coast of Japan. *Niijima* means "new island" in Japanese.

HAWAII, INDONESIA, THE GALÁPAGOS, ICELAND, AND NIIJIMA ARE EXAMPLES OF ISLANDS FORMED BY SUBMARINE VOLCANOES.

Even when they're not creating new islands, submarine volcanoes are changing the conditions in the ocean. Underwater vents and eruptions can release gases and magma—the molten or melted natural material from which lava is formed—that heat the water around them, creating columns of warm water. Different plant and animal life flock to these warmer temperatures. These warmer columns can also disrupt ocean currents; what once flowed one way must now flow in a slightly different direction. Additionally, gases released during an underwater eruption can alter the chemical makeup of the water. Some gases, for example, make the water a little more acidic, which may affect which plants and animals live near there. Other gases add nutrients to the water that can help plankton grow, which attracts larger fish and sea creatures.

Understanding how this all happens has driven Kenna Rubin's research.

She and other scientists are working hard to learn how volcanoes change over time and how magma is made deep beneath the ocean floor. They also want to better understand how these underwater occurrences affect our oceans.

"There's so much we still don't know," says Rubin. "It's like a giant puzzle. The volcanoes we observe, the rocks we're able to collect, the water samples we test—those are all pieces. We need to collect and fit together more pieces so we can begin to understand the bigger picture."

Exactly how does one learn about something located four miles beneath the ocean's surface? It's not easy.

First, researchers must determine where volcanic activity is taking place. They use machines called seismometers to detect shaking and other ground movement. They also compare high-resolution maps of the ocean floor over time. Is the landscape of a particular area changing month to month or year to year? They test water samples. Do they contain chemicals known to come out of erupting volcanoes?

If enough signals point to volcanic activity taking place, researchers may try to get a closer look. A lot of times, they'll use a remotely operated vehicle, or ROV. These high-tech devices are operated by pilots aboard a research ship, while a lead scientist provides directions. The ROVs can gather water samples, take photographs, and send live video to the ship. They also have arms that can collect rocks and other specimens.

"The ROVs are a lot like a video game," says Rubin. "Everything you're seeing is on a screen and the pilots use controls to move the ROV this way or that."

Rubin has been involved with hundreds of ROV-centered research trips. Given a choice, though, she much prefers studying submarine volcanoes using a human-occupied vehicle, or HOV.

HOVs are submersibles—like tiny submarines—that can carry two to three people onto the seafloor. This allows scientists to see things firsthand. These vehicles are small, with inside cabin space measuring just six or seven feet across. That may seem like plenty of room for a pilot and one or two researchers, but that space also contains diving gear, recording devices, computers, and more.

"It's very cozy, with everyone's legs touching everyone else's legs," said Rubin. "But I never think about the closeness or the fact that there's no bathroom, or even the fact that we're going to be sealed in this small space for ten hours.

"For me, the excitement of getting to see what so few people in the world get to see—that is so much cooler than a little physical discomfort. It's a true gift to experience what I've been able to experience."

Rubin says she often encounters students or other researchers who say they're simply not brave enough to travel to the bottom of the ocean in an HOV.

"These vehicles are very safe," she assures. "Plus, for me, it's not a matter of bravery, it's more a matter of curiosity. And as long as I've been studying volcanoes, I still have questions I'd love to be able to answer."

So, for now, Rubin and her fellow researchers will keep monitoring the volcanoes most people will never see. And they'll keep looking for answers to questions about the ever-changing landscape of the ocean floor.

CHAPTER 18
TREASURE HUNTERS

SHIPS HAVE BEEN CONNECTING THE WORLD FOR CENTURIES.

Many years before commercial airplanes were even imagined, ocean trade routes allowed ships to travel hundreds of miles in just a few days. People were able to move from place to place. Plus, these ships had room to carry cargo—crates and barrels packed with everything from fancy porcelain dishes and fragrant spices to gold coins and sparkling jewels.

Unfortunately, not every ship arrived at its destination.

Over the years, ships have collided with rocks, reefs, icebergs, and other boats. They've been tossed about by storms and taken on water. Wooden ships have gone up in flames. And direct hits from cannons or torpedoes have destroyed other vessels. No matter the cause, many of these ships, once full of high hopes and valuable cargo, have sunk to the bottom of the ocean.

How many shipwrecks are hidden in the briny depths? The Global Maritime Wrecks Database has records of more than 250,000 sunken vessels. But those are only the ones they know about.

Experts from UNESCO—the United Nations Educational, Scientific, and Cultural Organization—believe the remains of more than three million ships may be resting on the ocean floor. Three million!

The thought of finding millions—or billions—of dollars' worth of gold and gems in these long-lost ships has driven many modern-day treasure hunters to put on their scuba gear and start searching. The smartest hunters do their research, often homing in on the last known location of a specific ship or a once-popular trade route. They rent or buy high-tech devices. They hire crews and invest years in their underwater searches. But what happens if they *do* find one of these old ships? Who does the loot really belong to?

That's the $20 billion question that emerged with the discovery of the wreck of the *San José*.

A three-masted ship, the *San José* was an important part of the Spanish Navy's fleet. Its mission was to gather treasures from Spanish territories in the Americas and Caribbean to take back to Spain. The *San José*'s cargo of precious metals and emeralds was

> **EXPERTS FROM UNESCO BELIEVE THE REMAINS OF MORE THAN THREE MILLION SHIPS MAY BE RESTING ON THE OCEAN FLOOR.**

so valuable that she was usually protected by a fleet of warships to help ensure safe passage.

In 1708, however, Spain was at war with England, and the *San José*'s protection fleet was delayed. So she and her crew attempted to sail from Panama to Colombia on their own. And that's when the attack occurred.

Four English warships fired at the *San José*. Just as British soldiers prepared to board the vessel, an explosion ripped through the *San José*. The ship sank to the bottom of the ocean, taking nearly 600 crew members and 200 tons of treasure with it.

For almost 300 years, no one knew the exact location of the shipwreck. Then, in 1981, an American treasure-hunting company claimed that they had done what no one else could: They found the wreckage off the coast of Colombia—riches and all. Finders keepers! The loot was theirs! Except it wasn't that clear-cut—not at all.

When a shipwreck is found, many people want to claim the treasure. Some believe the goods should go to the people who discover it. Others say it belongs to the descendants of the ship's crew, descendants of the ship's owners, or the country from which the treasures were taken. Or, in the case of a naval ship, perhaps it should be returned to its home country.

So many questions, and they all bubbled to the surface with the discovery of the *San José*.

In spring 2024, even as multiple countries and private companies continued to battle over who should get the loot, the Colombian government planned to begin recovering the treasure. Colombian leaders insisted it wasn't about the money. They wanted to recover the ship's cargo so it could be displayed in museums to educate people about the country's history and culture.

The ultimate fate of the *San Jose*'s treasure is still being decided, and it is hardly the only shipwreck to have been discovered.

Long ago, most shipwrecks were found either by accident, by fishermen whose nets were caught on them, or by really persistent treasure hunters. Now, with the help of sonar equipment, high-tech submersibles, and high-definition cameras, finding deeper shipwrecks is becoming more commonplace.

In 2022, for example, Texas adventurer Victor Vescovo and his team found the deepest shipwreck ever identified. Using a deep-diving submersible, they discovered the USS *Samuel B. Roberts* more than four miles beneath the water's surface. The navy destroyer was resting in two pieces on a slope off the coast of the Philippines. Vescovo also identified what was previously the world's deepest wreck, the USS *Johnston,* in the same area in 2021.

WITH THE HELP OF SONAR EQUIPMENT, HIGH-TECH SUBMERSIBLES, AND HIGH-DEFINITION CAMERAS, **FINDING DEEPER SHIPWRECKS IS BECOMING MORE COMMONPLACE.**

A company was using a remote-operated vehicle to search for lost cargo containers off the coast of Australia when it located the wreck of the SS *Nemesis*. The steamship was transporting coal in 1904 when it vanished in a powerful storm with thirty-two people on board.

In 2024, a submersible robot operated by a natural gas company detected a 3,300-year-old ship at the bottom of the Mediterranean Sea. It is one of the oldest shipwrecks ever discovered. The ship was located off the coast of Israel, in water about a mile deep, with hundreds of intact jars still on board.

As enticing as treasure might be, shipwrecks are also considered great historical finds. Wrecks in shallow, warmer waters decay quickly, thanks to marine life and exposure to sunlight. But vessels found in deep, cold water with low oxygen levels are often well preserved and may contain important clues about a ship's crew, passengers, and cargo.

Archaeologists search for and study artifacts so they can learn about past civilizations. Underwater archaeologists, also called nautical, marine, or maritime archaeologists, have the same broad mission, but they do much of

> WRECKS IN SHALLOW, WARMER WATERS **DECAY QUICKLY**, THANKS TO MARINE LIFE AND EXPOSURE TO SUNLIGHT. BUT VESSELS FOUND IN DEEP, COLD WATER WITH LOW OXYGEN LEVELS ARE OFTEN **WELL PRESERVED.**

their work, as you might guess, underwater. They search for artifacts located in bodies of water that might provide information about the past. They want to learn how people throughout history and from lost cultures have interacted with oceans, seas, rivers, and lakes. Using remote sensors, scuba gear, metal detectors, underwater cameras, and excavation, these experts can piece together information about ancient trade routes and technologies, as well as the lives and traditions of the people and communities who lived near the water.

UNDERWATER ARCHAEOLOGISTS, ALSO CALLED NAUTICAL, MARINE, OR MARITIME ARCHAEOLOGISTS, WANT TO LEARN HOW PEOPLE THROUGHOUT HISTORY AND FROM LOST CULTURES HAVE INTERACTED WITH OCEANS, SEAS, RIVERS, AND LAKES.

Marine archaeologists are often supported by universities and nonprofit groups. The Institute of Nautical Archaeology, for example, has assisted at many sites, including a 3,000-year-old shipwreck off the coast of Türkiye, as well as the first Phoenician shipwreck to be excavated near Cartagena, Spain. Another organization, Diving with a Purpose, has more than 500 volunteer divers. Its main goal is to work with marine archaeologists to locate and document shipwrecks related to the transatlantic slave trade.

Some treasure hunters dig through shipwrecks or use dynamite to blow holes in rusty bulkheads or decks—whatever it takes to find the loot that will make them rich. These destructive actions put them at odds with the archaeologists who treat shipwrecks like crime scenes, meticulously sifting through and documenting clues. Human bodies are rarely found in ancient shipwrecks; they've usually been swept away or eaten by fish. But well-preserved skeletons may be found, and archaeologists argue that they are the ones best prepared to handle those remains with care and reverence.

Battles over shipwrecks have ended up in court, with judges and juries sometimes taking years to make decisions. Treasure or knowledge? The oceans hold both, plus millions more mysteries waiting to be uncovered.

CHAPTER 19
UNDERWATER CITIES

UNDERWATER CITIES HAVE PROVIDED THE SETTING FOR countless movies, shows, novels, and video games: Atlantis, Otoh Gunga, even Bikini Bottom. These fictional hidden worlds, inhabited by remarkable civilizations and monstrous creatures, are about as high drama as you can get.

But—news flash—there are *real* underwater cities. And not just one or two. There are at least fifteen cities worldwide that once sat on dry land and are now located at the bottom of the ocean.

How does a city sink? Well, there are a number of ways this could happen.

Floods are responsible for submerging most of these cities. Rising sea levels, earthquakes, tsunamis, and landslides also get some of the blame. No matter the reason, these previously thriving cities are now completely underwater. According to the

United Nations Educational, Scientific, and Cultural Organization (UNESCO), submerged settlements have been found in waters near Egypt, England, India, Jamaica, Argentina, Denmark, Sweden, and Italy and in the Black Sea.

Baiae, also known as Baia, was once a vibrant Roman city. It was where the rich and famous vacationed. Poets traveled there for inspiration. Powerful politicians built fancy houses, complete with spas and pools. Its location, along the Gulf of Naples and on the slopes of a supervolcano called Campi Flegrei, was both its appeal and its undoing.

A phenomenon called bradyseism, which is Greek for "slow movement," was causing the ground around Baiae to gradually rise and fall. These shifts are the result of thousands of minor earthquakes and magma shifting deep under Earth's surface. It's something that often happens near volcanoes, but geologists and city planners hadn't exactly figured that out way back then. As early as the third century CE, the ground around Baiae began to sink into the sea. Bit by bit, the water swallowed its villas and temples. By the eighth century, the lower part of Baiae was mostly submerged.

DUE TO A PHENOMENON CALLED **BRADYSEISM**, WHICH IS GREEK FOR "SLOW MOVEMENT," THE GROUND AROUND BAIAE BEGAN TO SINK INTO THE SEA. BY THE EIGHTH CENTURY, THE LOWER PART OF BAIAE WAS MOSTLY SUBMERGED.

These days Baiae's remains are featured at one of the world's few underwater archeological parks. Well-preserved statues, ruins of ancient shops and villas, crumbling columns, mosaic floors, and more can be seen by scuba divers or visitors in glass-bottom boats.

The town of Port Royal, Jamaica, experienced a much more sudden shift from aboveground to below the water's surface. Located at the mouth of Kingston Harbor, Port Royal was founded in 1494. It quickly emerged as a center for commerce and shipping. Then, in the mid-1600s, Port Royal earned a reputation as a famous hideout for criminals. Looting, smuggling, fighting, and drinking were commonplace. Buccaneers and pirates, including Blackbeard and Calico Jack, hung out here. It became known as "the wickedest city on earth."

More than 6,500 people were living in Port Royal in 1692. That's when a powerful earthquake devastated the town. The quake's tremors triggered a tsunami, or giant wave, to strike Port Royal's shores. Those forces worked together to liquefy the ground beneath the town. About two-thirds of the buildings slid right into the ocean.

Port Royal's minister, Rev. Edmund Heath, survived the earthquake and found safety on a ship moored in the harbor. As he looked out upon the ruins of the city, he wrote: "The earth opened and swallowed many people, before my face, and the sea I saw came mounting in over the wall, upon which I concluded it impossible to escape."

Today, most of the remains of Port Royal lie under 40 feet (12 m) of water. Special access is required to dive in the area. Since the 1950s, researchers have been exploring and cataloging the underwater town. Port Royal is a designated National Heritage Site, and it's considered one of the most important underwater archaeological sites in the Western Hemisphere. Many of the items recovered from the submerged city are now on display at the Museums of History and Ethnography at the Institute of Jamaica in Kingston. One of those items is a pocket watch that stopped at the exact time and date of the earthquake, 11:43 a.m. on June 7, 1692.

> **PORT ROYAL** IS A DESIGNATED NATIONAL HERITAGE SITE, AND IT'S CONSIDERED ONE OF THE MOST IMPORTANT UNDERWATER ARCHAEOLOGICAL SITES IN THE WESTERN HEMISPHERE.

Heracleion, Egypt, also known as Thonis, was once a hustling, bustling trade center. Canals crisscrossed the city, connecting houses, businesses, temples, and wharves. Using a network of ferries, bridges, and pontoons, most of the goods being shipped from the Mediterranean into Egypt came through Heracleion.

The city's demise was gradual. Archaeologists say it was destroyed by a combination of earthquakes, tsunamis, and rising sea levels over many years, with the last of its buildings collapsing into the water in the eighth century CE.

In 1933, a pilot flying over the area saw ruins in the water. His reported sighting got others excited about mapping the area and searching for Heracleion.

The lost city was found again in 2000, just 30 feet (9 m) underwater. Marine archaeologists have recovered artifacts including giant statues, pottery, jewelry, and coins. Research at the site is ongoing.

Natural disasters aren't to blame for all sunken cities. Shi Cheng, China, sits not on the ocean floor but, rather, on the bottom of a lake. And how it got there was no accident.

This 1,400-year-old city, surrounded by city gates and towers, covered an area roughly equal to sixty football fields. Shi Cheng, also known as Lion City, had wide paved roads. Its buildings were adorned with stone statues of lions, dragons, and phoenixes. The city, located at the foot of Wu Shi Mountain, was once the center of politics in Zhejiang, one of China's eastern coastal provinces.

In 1959, the 300,000 people who lived in Shi Cheng were told to pack up and move. All of them. The residents were forced to relocate so the city could be intentionally flooded to make way for a new dam and hydroelectric station. Shi Cheng now sits 130 feet (40 m) under the surface of Qiandao Lake.

IN 1959, THE 300,000 RESIDENTS OF SHI CHENG WERE FORCED TO RELOCATE SO THE CITY COULD BE **INTENTIONALLY FLOODED** TO MAKE WAY FOR A NEW DAM AND HYDROELECTRIC STATION.

The city was all but forgotten until 2001, when the Chinese government organized an expedition to see if any of its buildings were still intact. They were.

Shi Cheng's buildings, carvings, and arches had been well preserved thanks to the lake's dark waters. Divers and researchers are still working to photograph and document items found there, including pottery, bronze ware, and stone tablets.

Could a modern-day city find itself at the bottom of an ocean, sea, or lake?

Experts say it's not only possible, but likely. Sea levels are on the rise, and global warming is largely responsible. The encroaching water is coming from melting ice sheets and glaciers, and the expansion of seawater as it warms. Low-lying coastal cities are already experiencing devastating floods.

Many regions are pumping so much water out of the ground that they are causing land to sink. That's what's happening in Jakarta, Indonesia. That city is sinking at a rate of more than 6 inches (15 cm) per year. Unless changes are made quickly, much of the city could be underwater by 2050.

Lagos, Nigeria, is Africa's largest city. Rising seas put it in danger of flooding. Houston, Texas, is

THE CITY OF **JAKARTA IS SINKING** AT A RATE OF MORE THAN 6 INCHES (15 CM) PER YEAR. UNLESS CHANGES ARE MADE QUICKLY, MUCH OF THE CITY COULD BE UNDERWATER BY 2050.

sinking at a rate of 2 inches (5 cm) per year, which makes it more susceptible to damage from hurricanes and floods. Land-use specialists, climate scientists, and data analysts predict other cities may also be at risk, including Dhaka, Bangladesh; Venice, Italy; Bangkok, Thailand; Rotterdam, the Netherlands; and Alexandria, Egypt. Even Virginia Beach, New Orleans, and Miami in the United States are at risk.

Human activities like burning fossil fuels, manufacturing products, cutting down forests, and farming livestock contribute to increased greenhouse gases in the atmosphere, causing the planet to warm even more quickly, furthering the rising sea levels.

Experts say even if the world completely stopped emitting greenhouse gases tomorrow—as if that were possible—ocean levels would continue to rise. Is it a lost cause? No way, says Paul Kirchen, an engineer specializing in climate change adaptations.

"Fortunately, many very intelligent and dedicated people in the government, in universities, in businesses, and in communities understand the problem well and are working hard to determine ways to lessen the impacts," says Kirchen, who teaches at the

HUMAN ACTIVITIES LIKE BURNING FOSSIL FUELS, MANUFACTURING PRODUCTS, CUTTING DOWN FORESTS, AND FARMING LIVESTOCK **CONTRIBUTE TO THE RISING SEA LEVELS.**

University of Massachusetts Boston. "They know the solutions and are trying to figure out how to put them in place."

He says ensuring more cities don't slip into the ocean will require companies, governments, and communities all over the world to work together. "It is important that young people support these efforts by learning about the problems," he said. "They can share their concerns with their families and friends. And some may become experts in managing these problems as they get older."

CHAPTER 20
A WHALE OF A TALE

IT WAS A CLEAR FIFTY-DEGREE MORNING ON JUNE 11, 2021, when commercial lobster diver Michael Packard set off for work. Like most mornings, Packard met crewmate Josiah Mayo on his boat, the *Ja'n J.*

Most lobster fishermen use traps baited with herring to catch the crustaceans. Packard, however, prefers to suit up in scuba gear and dive to the ocean floor. Wearing heavy gloves, he grabs the live lobsters and plunks them into a bag, which he then carries to the surface of the water.

On this June morning, Packard and Mayo had anchored their boat off Herring Cove Beach, in Cape Cod Bay. They were surrounded by other boats whose crews were fishing for striped bass, bluefin tuna, and butterfish. Packard had already made two successful thirty-minute dives and had loaded bags filled with more than 100 pounds (45 kg) of lobsters onto the boat.

Dive number three was when everything changed. Packard was about 10 feet (3 m) from the ocean floor when he was slammed. Hard.

"Just like a freight train. Just boom," he later told *The Cape Cod Times.* "And then, all the sudden, it went black. And water was just rushing—rushing around me and black and I could feel pressure on my whole body. And I was just moving through the water wicked fast."

Shark. That was Packard's first thought. He had been attacked and swallowed by a shark. After all, he often saw them when he dove. But it didn't take him long to realize he didn't feel any teeth. Plus, wherever he was, well, it seemed larger than a shark.

And that's when it clicked. *He was in a whale's mouth.* It wasn't a place he'd ever imagined being.

Packard could feel the whale squeezing him. His regulator—the device that provided air from his scuba tank—had been jerked out of his mouth. Once he got that back in place, he decided he didn't have a lot of options except to fight.

"I was struggling and banging and kicking and just thinking there's no way I'm gonna get out of this unless he decides to let me go," Packard said. The Massachusetts man worried the whale might dive even deeper in search of food. Though he feared for his life, Packard didn't give up. He punched and kicked harder.

At the same time Packard was fighting for his life, Mayo started to sense something was wrong. He could see that the bubbles from Packard's regulator were moving away from the boat. There was also an abnormal amount of splashing and splattering going on. It was almost as if the water had come to a boil. Mayo worried it might be a shark. After all, great white sharks are drawn to Cape Cod because of its growing seal population.

"Then I saw whale parts," the crewmate recalled. "I saw whale fluke and the whale's head, and I suddenly knew it wasn't a shark attack."

To Packard, the underwater struggle felt like it lasted an eternity. In reality, it was probably only thirty to forty seconds before the whale swam to the surface of the water. And then, as quickly as it had swallowed Packard, the whale spit him back out.

GREAT WHITE SHARKS ARE DRAWN TO CAPE COD BECAUSE OF ITS GROWING SEAL POPULATION.

"I saw light and he started throwing his head side to side," said Packard. "And the next thing I knew I was outside," in the water.

Mayo and another fisherman were able to pull Packard into the boat. They called for an ambulance and hurried to meet it at a nearby pier. At first, Packard thought the whale might have broken his legs. In the end, he was left with a sprained knee, some deep bruising on his upper legs, and a survival story for the ages.

Many people were initially skeptical of the whole story. Decompression sickness, also known as the bends, occurs when a diver ascends too quickly. At least one doctor who treated Packard told reporters he wondered how Packard could rise so quickly through the water—even if he was inside a whale—and not get sick. Other whale experts chimed in claiming the story sounded a little fishy.

Still, eyewitness reports and Packard's injuries added credibility to the tale. More experts spoke up. The whale Packard encountered was a humpback, most likely a juvenile. Humpbacks, which average around fifty feet long, are a common sight off the coast of New England. They are not aggressive toward humans, but that doesn't mean the encounter didn't happen.

Jooke Robbins, director of Humpback Whale Studies at the Center for Coastal Studies in Provincetown, Massachusetts, suggested the whale in question might have been feeding on small wiggly fish called sand lance. When humpbacks eat, they often plow through concentrated areas of food with their mouths wide open. This allows them to gulp enormous mouthfuls of prey and seawater. This whale's particular

> WHEN **HUMPBACKS EAT**, THEY OFTEN PLOW THROUGH CONCENTRATED AREAS OF FOOD **WITH THEIR MOUTHS WIDE OPEN**.

gulp happened to include Packard. Baleen plates hang from the whale's upper jaw. These stiff, bristly plates act like a filter or sieve;

the prey stays in the whale's mouth while the water strains out. At some point the whale must have realized he'd taken in something too large—and too feisty—to swallow.

"When a humpback moves to lunge, its mouth is open hugely, and its eyes, they bulge out a little bit, but they're off to the sides," Robbins told *The Provincetown Independent*. "It's quite likely that it can't see right in front of its face when it's feeding like that. Based on our work with whale disentanglement, it's quite clear that, when they're feeding, they're getting things in their mouth they didn't want. Usually, it's fishing gear. This time, it was Mike."

> "BASED ON OUR WORK WITH WHALE DISENTANGLEMENT, IT'S QUITE CLEAR THAT, WHEN THEY'RE FEEDING, THEY'RE GETTING THINGS IN THEIR MOUTH THEY DIDN'T WANT. USUALLY, IT'S FISHING GEAR. **THIS TIME, IT WAS MIKE.**"
>
> **—JOOKE ROBBINS,** DIRECTOR OF HUMPBACK WHALE STUDIES AT THE CENTER FOR COASTAL STUDIES

As Packard recounts the event, he is quick to point out he was in the whale's mouth, but he was not swallowed by it. The inside of a humpback's mouth can be the size of a small car. But its throat can only stretch to about 15 inches (38 cm). Gulped. Not swallowed. Accuracy matters.

Still, the story is sensational, and it was repeated by news outlets around the world. Packard even got an invitation to tell his tale on the late-night TV show *Jimmy Kimmel Live!* Kimmel suggested that perhaps the lobsters had teamed up with the whale.

"Did you sense that the lobsters were watching and felt like they were getting some kind of revenge?" he joked.

Packard laughed along, agreeing that the incident had been an unlucky one for both him and the whale. He also issued a heartfelt apology to the whale "for getting in his way. And I won't ever do it again."

SELECTED SOURCES

I HAVE ALWAYS BEEN DRAWN TO STORIES THAT ARE SO WILD and unexpected that my immediate thought is: *Well, that can't be real.* And then—hold on to your hat—IT IS TRUE! I call them mindblowers, and it's been a delight to research and gather this collection of unbelievable-but-true stories about our oceans and seas. Research for a book like this is much like the ocean itself: vast and far-reaching. But hands-on investigation has always been my favorite, so of course, I let plenty of chilly waves wash over my toes, waded out to explore a concrete ship's ruins, and convinced an octopus wrangler to let me feel the *pop, pop, pop* of the creature's suction cups pulling off my skin. Interviews with scientists, activists, and adventurers added another layer of depth to these stories, as did books and articles. Below is a selection of the people, organizations, websites, and publications consulted for this project.

WEBSITES

The **AMERICAN MUSEUM OF NATURAL HISTORY** hosts a page on their website called **OLOGY**, meaning "a branch of knowledge," where you can find games, videos, quizzes, and even hidden trading cards with facts about marine biology and other areas of study.

amnh.org/explore/ology/marine-biology

The **OCEAN PORTAL** from **NATIONAL GEOGRAPHIC KIDS** is filled with puzzles, fun facts, and plenty of information about the ways in which kids can help protect Earth's oceans.

kids.nationalgeographic.com/pages/topic/ocean-portal

The **NATIONAL OCEANIC AND ATMOSPHERIC ADMINISTRATION'S NATIONAL OCEAN SERVICE** division works to monitor ocean conditions, conduct deep-sea exploration, and protect marine mammals and endangered species in the United States.

oceanservice.noaa.gov

THE EXPLODING WHALE is a volunteer-run website dedicated to documenting Oregon's most famous beached whale. It includes interviews, photos, and even videos of the event.

theexplodingwhale.com

BOOKS

Brown, Aly. *The Last Explored Place on Earth: Investigating the Ocean Floor with Alvin the Submersible.* Feiwel and Friends, 2023.

Callahan, Steven. *Adrift: Seventy-Six Days Lost at Sea*. Houghton Mifflin Harcourt, 1986.

Cusolito, Michelle. *A Window into the Ocean Twilight Zone: Twenty-Four Days of Science at Sea*. Charlesbridge, 2024.

Ebbesmeyer, Curtis, and Eric Scigliano. *Flotsametrics and the Floating World: How One Man's Obsession with Runaway Sneakers and Rubber Ducks Revolutionized Ocean Science*. Smithsonian Books, 2009.

Eriksen, Marcus. *Junk Raft: An Ocean Voyage and a Rising Tide of Activism to Fight Plastic Pollution*. Beacon Press, 2017.

Hohn, Donovan. *Moby-Duck: The True Story of 28,800 Bath Toys Lost at Sea and of the Beachcombers, Oceanographers, Environmentalists, and Fools, Including the Author, Who Went in Search of Them*. Viking Press, 2011.

Kemp, Christopher. *Floating Gold: A Natural (and Unnatural) History of Ambergris*. University of Chicago Press, 2012.

Newman, Patricia. *Planet Ocean: Why We All Need a Healthy Ocean*. Millbrook Press, 2021.

Nickum, Nora. *Superpod: Saving the Endangered Orcas of the Pacific Northwest*. Chicago Review Press, 2023.

Recio, Belinda. *When Animals Rescue: Amazing True Stories About Heroic and Helpful Creatures*. Skyhorse Publishing, 2021.

Sandler, Martin W. *Shipwrecked! Diving for Hidden Time Capsules on the Ocean Floor*. Astra Young Readers, 2023.

ORGANIZATIONS AND INDIVIDUALS

MARCUS ERIKSEN and **ANNA CUMMINS** are cofounders of the **5 GYRES INSTITUTE**. Their team partners with researchers, corporations, and communities to continue investigating plastic pollution in our waters and solutions to combat its impact.

5gyres.org

STEPHEN P. LEATHERMAN, AKA DR. BEACH, is director of the Laboratory for Coastal Research at Florida International University and one of the world's foremost beach experts.

drbeach.org

MARINEBIO CONSERVATION SOCIETY is a nonprofit organization committed to educating others about sea life and the value of ocean conservation. On MarineBio Kids, learn about the day-to-day work of a marine biologist, solve puzzles, and take online classes created just for young people.

marinebio.org/kids

The **OCEAN CONSERVANCY** focuses on threats to oceans and seas. The website includes information about dozens of marine animals and birds.

oceanconservancy.org

ORCA NETWORK is an organization in the Pacific Northwest that connects whales and people through a learning center, hosting a whale-sighting network and sharing videos of southern resident orcas, among other initiatives.

orcanetwork.org

The **SCHMIDT OCEAN INSTITUTE** organizes research cruises and ROV (remotely operated vehicle) submarine exploration ventures along with other projects in an effort to advance research, discovery, and knowledge related to oceans.

schmidtocean.org

The **WOODS HOLE OCEANOGRAPHIC INSTITUTION** is an independent nonprofit organization dedicated to advancing science and engineering in the name of ocean research, exploration, and education.

whoi.edu

GLOSSARY

A

AMBERGRIS: a waxy substance excreted by sperm whales that is used to make perfumes

APEX PREDATOR: a predator at the top of a food chain that is not preyed upon by any other animal

ARCTIC PACK ICE: pieces of ice that collided to form a large mass in the Arctic Ocean and nearby waterways. Pack ice is not attached to land.

AUTONOMOUS UNDERWATER VEHICLE (AUV): a robot submarine that explores the ocean without a pilot or a tether

B

BALEEN PLATE: a hairlike structure in the mouth of a baleen whale that filters food from seawater. Baleen plates are made of keratin, the same protein that makes up human hair and fingernails.

BEACHING: when marine life such as a whale or a dolphin becomes stranded on the shore or a beach

BRADYSEISM: the gradual lifting or falling movement of Earth's crust

BREAKWATER: an offshore structure that protects a harbor or beach from the force of waves

BUBONIC PLAGUE: an infectious bacteria spread primarily through fleas or rats that was prevalent during the Middle Ages

BUOYANCY: an object's ability to float

C

CARBON DATING: a scientific method of calculating the age of organic matter by measuring the amount of carbon inside it

CEPHALOPOD: member of the molluscan class that includes squid, cuttlefish, and octopuses and is characterized by highly developed eyes, heads located directly above their limbs, and at least eight arms or tentacles

CHITIN: a natural substance found in the exoskeletons of insects and the shells of crustaceans

CONSERVATION: an effort to use natural resources wisely with the intention of protecting the environment to ensure that it can continue to support life

CONSERVATION CANINE: a specially trained dog used to sniff odors related to conservation, such as the scat of endangered species, invasive plant life, or environmental contaminants

CRUSTACEAN: an animal that usually has a hard covering or exoskeleton and two pairs of antennae

D

DISPLACEMENT: the change in position of an object, such as the movement of water when an object is submerged

DNA: genetic material found in the cells of every living thing that determines how an organism will look and function

E

EDDY: a current of air or water running against the main current or in a circle

F

FERRO-CEMENT: a building material made of thin cement slabs reinforced with steel mesh

FLOTSAM: debris in the water that was not deliberately thrown overboard, often as the result of a shipwreck

FREE DIVE: to swim beneath the surface of water without a portable breathing device

G

GYRE: a large circular system of ocean currents that rotates around a central point

H

HADOPELAGIC ZONE: also known as the hadal zone; the deepest, coldest, and darkest region of the ocean

HUMAN OCCUPIED VEHICLE (HOV): carries researchers and equipment onto the seafloor, allowing in-person research and observation

M

MAGMA: hot, liquefied rock that rises to the top of a volcano

MARINE BIOLOGIST: a scientist who studies the behaviors and environments of plants and animals that live in or near the water

MATRILINEAL: parentage that can be tracked through the mother's line of succession. In matrilineal societies, relationships center around the mother.

MICROBEAD: a tiny piece of plastic that can be found in personal care products such as cleansers and toothpastes

MICROPLASTIC: plastic pieces less than 5 millimeters long that can be harmful to marine life

MOLLUSK: a large group of animals with soft bodies and no spines

N

NOCTURNAL: a living thing that sleeps during the day and becomes active during the night

P

PASSIVE LAVA FLOW: type of volcanic eruption where molten rock, or lava, gently flows from the vent of the volcano, causing a nonexplosive eruption

PLANKTON: tiny living things, including algae and bacteria, that float and drift in oceans and other bodies of water

R

REMOTELY OPERATED VEHICLE (ROV): an unoccupied underwater machine operated by someone at the water's surface to explore the ocean's depths

RESIDENT ORCA: a fish-eating orca that lives in large pods and stays within a family group its entire life

RIP CURRENT: a powerful, narrow channel of water that flows quickly away from shore

S

SEAMOUNT: an underwater mountain

SEISMOGRAPH: a device used to record the strength and duration of an earthquake

SEISMOMETER: the internal mechanism of a seismograph that measures and records movement during an earthquake

SUBMARINE VOLCANO: a volcano that forms underwater

SUSPENSION BRIDGE: a bridge suspended from cables anchored at either end and usually supported by towers

T

TERRESTRIAL VOLCANO: a volcano on Earth's surface

TRANSDIFFERENTIATION: the conversion of one cell type to another and the process by which the *Turritopsis dohrnii*, or immortal jellyfish, regenerates through its life cycle

TSUNAMI: a series of large ocean waves generated by undersea disturbances such as earthquakes

U

UNDERWATER ARCHAEOLOGIST: a researcher who recovers and documents information from underwater sites, often shipwrecks, to learn about past human civilizations and practices

ACKNOWLEDGMENTS

MY LOVE FOR STRANGE-BUT-TRUE FACTS DATES BACK TO THE fourth grade, when I routinely wrote unassigned reports with sensational titles like "I Spy Owls with No Eyeballs" and "Can You Hear That? The Trees Are Talking." My teacher, Ms. Helen Schmidt, could have laughed and sent me on my way. Instead, she noted her favorite passages and encouraged my curiosity, research, and writing. Sharing some of my early published books with her remains one of the greatest joys of my life. My first and deepest thanks go to Ms. Schmidt and to educators everywhere who inspire a sense of wonder.

The research for this book was made possible by the support of many librarians, scientists, and conversationists. Thank you for your time, insights, and help uncovering the perfect facts to bring these stories to life. I am especially grateful to environmental activists and educators, including Marcus Eriksen and Anna Cummins of the 5 Gyres Institute, Monika Wieland Shields of the Orca Behavior Institute, and Jay Julius of the Lummi Nation. Your dedication and hard work remind us that the ocean is a shared treasure worth protecting.

Gratitude to the Rights Factory and my agent, Stacey Kondla, whose belief in me has been steadfast. Stacey's calm demeanor is the perfect balance to my perpetual impatience.

To my critique partners—Jessica Froberg, Adria Goetz, Jennifer Nielsen, Elizabeth Scherman, Steph Scott, and Paula VanEnkevort—and my walk-and-talk writing pal Jenny: Thank you for asking hard questions, pushing me to dig deeper, and keeping me on track when I got distracted along the way.

I'm immensely thankful to the entire team at Bright Matter Books and Penguin Random House for believing in this project. From the moment publisher Tom Russell and I first geeked out over our shared love of obscure facts, I knew this book was in the right hands. Having the opportunity to work with and learn from super editor Elizabeth Stranahan has made me a better writer. Copy editors Alison Kolani and Maddy Stone, managing editor Rebecca Vitkus, designer Jade Rector, cover and interior artist Max Temescu, and production manager Patty Collins are the dream team behind this book's shine.

Finally, to my husband, Mitch, and our kids, Eve and Eli: Thank you for enduring more than your share of dinner-table discussions about trash-laden tides and valuable whale poop. This book exists because of Team Boone-Rob's love, humor, and belief in me and my often-wacky ideas.

ABOUT THE AUTHOR

MARY BOONE grew up in Iowa and vividly recalls her family's first trip to Florida when she was seven years old. The vastness of the Atlantic Ocean took her breath away and ignited her curiosity. (Yes, it's possible to be breathless and curious at the same time!) A former newspaper reporter, Mary now lives with her family a short walk from the Puget Sound in Washington, where she writes nonfiction and informational fiction books for curious readers. Her previous books include *Bugs for Breakfast: How Eating Insects Could Help Save the Planet*, *School of Fish*, and *Pedal Pusher: How One Woman's Bicycle Adventure Helped Change the World*. To learn more about Mary, visit her website at boonewrites.com.